CALCULUS
SUCCESS
in 20 Minutes a Day

CALCULUS SUCCESS
in 20 Minutes a Day

Second Edition

Mark A. McKibben
Christopher Thomas

LEARNINGEXPRESS®

NEW YORK

Library of Congress Cataloging-in-Publication Data:
McKibben, Mark A.
  Calculus success in 20 minutes a day / Mark A. McKibben.—2nd ed.
      p.   cm.
  Previous ed.: Calculus success in 20 minutes a day / Thomas, Christopher. © 2006.
  ISBN 978-1-57685-889-9
  1. Calculus—Problems, exercises, etc.   I. Thomas, Christopher, 1973– Calculus success
in 20 minutes a day.   II. Title.   III. Title: Calculus success in twenty minutes a day.
  QA303.2.T47 2012
  515—dc23

                                                                            2011030506

Printed in the United States of America

9 8 7 6 5 4 3 2 1

ISBN 978-1-57685-889-9

For information or to place an order, contact LearningExpress at:
  2 Rector Street
  26th Floor
  New York, NY 10006

Or visit us at:
  www.learnatest.com

ABOUT THE AUTHOR

Mark A. McKibben is a professor of mathematics and computer science at Goucher College in Baltimore, Maryland. During his 12 years at this institution, he has taught more than 30 different courses spanning the mathematics curriculum, and has published two graduate-level books with CRC Press, more than two dozen journal articles on differential equations, and more than 20 supplements for undergraduate texts on algebra, trigonometry, statistics, and calculus.

Christopher Thomas is a professor of mathematics at the Massachusetts College of Liberal Arts. He has taught at Tufts University as a graduate student, Texas A&M University as a postdoctorate professor, and the Senior Secondary School of Mozano, Ghana, as a Peace Corps volunteer. His classroom assistant is a small teddy bear named e^x.

ABOUT THE AUTHOR

CONTENTS

INTRODUCTION ▶

If you have never taken a calculus course, and now find that you need to know calculus—this is the book for you. If you have already taken a calculus course, but felt like you never understood what the teacher was trying to tell you—this book can teach you what you need to know. If it has been a while since you have taken a calculus course, and you need to refresh your skills—this book will review the basics and reteach you the skills you may have forgotten. Whatever your reason for needing to know calculus, *Calculus Success in 20 Minutes a Day* will teach you what you need to know.

Overcoming Math Anxiety

Do you like math or do you find math an unpleasant experience? It is human nature for people to like what they are good at. Generally, people who dislike math have not had much success with math.

If you have struggles with math, ask yourself why. Was it because the class went too fast? Did you have a chance to fully understand a concept before you went on to a new one? One of the comments students frequently make is, "I was just starting to understand, and then the teacher went on to something new." That is why *Calculus Success* is self-paced. You work at your own pace. You go on to a new concept only when you are ready.

When you study the lessons in this book, the only person you have to answer to is you. You don't have to pretend you know something when you don't truly understand. You get to take the time you need to understand everything before you go on to the next lesson. You have truly learned something only when you thoroughly understand it. Take as much time as you need to understand examples. Check your work with the answers and if you don't feel confident that you fully understand the lesson, do it again. You might think you don't want to take the time to go back over something again; however, making sure you understand a lesson completely may save you time in the future lessons. Rework problems you missed to make sure you don't make the same mistakes again.

How to Use This Book

Calculus Success teaches basic calculus concepts in 20 self-paced lessons. The book includes a pretest, a posttest, 20 lessons, each covering a new topic, and a glossary. Before you begin Lesson 1, take the pretest. The pretest will assess your current calculus abilities. You'll find the answer key at the end of the pretest. Each answer includes the lesson number that the problem is testing. This will be helpful in determining your strengths and weaknesses. After taking the pretest, move on to Lesson 1, *Functions*.

Each lesson offers detailed explanations of a new concept. There are numerous examples with step-by-step solutions. As you proceed through a lesson, you will find tips and shortcuts that will help you learn a concept. Each new concept is followed by a practice set of problems. The answers to the practice problems are in an answer key located at the end of the book.

When you have completed all 20 lessons, take the posttest. The posttest has the same format as the pretest, but the questions are different. Compare the results of the posttest with the results of the pretest you took before you began Lesson 1. What are your strengths? Do you still have weak areas? Do you need to spend more time on some concepts, or are you ready to go to the next level?

Make a Commitment

Success does not come without effort. If you truly want to be successful, make a commitment to spend the time you need to improve your calculus skills.

So sharpen your pencil and get ready to begin the pretest!

PRETEST

Before you begin Lesson 1, you may want to get an idea of what you know and what you need to learn. The pretest will answer some of these questions for you. The pretest consists of 50 multiple-choice questions covering the topics in this book. While 50 questions can't cover every concept or skill taught in this book, your performance on the pretest will give you a good indication of your strengths and weaknesses.

If you score high on the pretest, you have a good foundation and should be able to work through the book quickly. If you score low on the pretest, don't despair. This book will explain the key calculus concepts, step by step. If you get a low score, you may need to take more than 20 minutes a day to work through a lesson. However, this is a self-paced program, so you can spend as much time on a lesson as you need. *You* decide when you fully comprehend the lesson and are ready to go on to the next one.

Take as much time as you need to complete the pretest. When you are finished, check your answers with the answer key at the end of the pretest. Along with each answer is a number that tells you which lesson of this book teaches you about the calculus skills needed to answer that question. You will find the level of difficulty increases as you work your way through the pretest.

1. ⓐ ⓑ ⓒ ⓓ
2. ⓐ ⓑ ⓒ ⓓ
3. ⓐ ⓑ ⓒ ⓓ
4. ⓐ ⓑ ⓒ ⓓ
5. ⓐ ⓑ ⓒ ⓓ
6. ⓐ ⓑ ⓒ ⓓ
7. ⓐ ⓑ ⓒ ⓓ
8. ⓐ ⓑ ⓒ ⓓ
9. ⓐ ⓑ ⓒ ⓓ
10. ⓐ ⓑ ⓒ ⓓ
11. ⓐ ⓑ ⓒ ⓓ
12. ⓐ ⓑ ⓒ ⓓ
13. ⓐ ⓑ ⓒ ⓓ
14. ⓐ ⓑ ⓒ ⓓ
15. ⓐ ⓑ ⓒ ⓓ
16. ⓐ ⓑ ⓒ ⓓ
17. ⓐ ⓑ ⓒ ⓓ

18. ⓐ ⓑ ⓒ ⓓ
19. ⓐ ⓑ ⓒ ⓓ
20. ⓐ ⓑ ⓒ ⓓ
21. ⓐ ⓑ ⓒ ⓓ
22. ⓐ ⓑ ⓒ ⓓ
23. ⓐ ⓑ ⓒ ⓓ
24. ⓐ ⓑ ⓒ ⓓ
25. ⓐ ⓑ ⓒ ⓓ
26. ⓐ ⓑ ⓒ ⓓ
27. ⓐ ⓑ ⓒ ⓓ
28. ⓐ ⓑ ⓒ ⓓ
29. ⓐ ⓑ ⓒ ⓓ
30. ⓐ ⓑ ⓒ ⓓ
31. ⓐ ⓑ ⓒ ⓓ
32. ⓐ ⓑ ⓒ ⓓ
33. ⓐ ⓑ ⓒ ⓓ
34. ⓐ ⓑ ⓒ ⓓ

35. ⓐ ⓑ ⓒ ⓓ
36. ⓐ ⓑ ⓒ ⓓ
37. ⓐ ⓑ ⓒ ⓓ
38. ⓐ ⓑ ⓒ ⓓ
39. ⓐ ⓑ ⓒ ⓓ
40. ⓐ ⓑ ⓒ ⓓ
41. ⓐ ⓑ ⓒ ⓓ
42. ⓐ ⓑ ⓒ ⓓ
43. ⓐ ⓑ ⓒ ⓓ
44. ⓐ ⓑ ⓒ ⓓ
45. ⓐ ⓑ ⓒ ⓓ
46. ⓐ ⓑ ⓒ ⓓ
47. ⓐ ⓑ ⓒ ⓓ
48. ⓐ ⓑ ⓒ ⓓ
49. ⓐ ⓑ ⓒ ⓓ
50. ⓐ ⓑ ⓒ ⓓ

1. What is the value of $f(4)$ when $f(x) = 3x^2 - \sqrt{x}$?
   a. 22
   b. 46
   c. 140
   d. 142

2. Simplify $g(x + 3)$ when $g(x) = x^2 - 2x + 1$.
   a. $x^2 + 2x + 4$
   b. $x^2 - 2x + 4$
   c. $x^2 + 4x + 16$
   d. $x^2 + 4x + 13$

3. What is $(f \circ g)(x)$ when $f(x) = x - \dfrac{2}{x}$ and $g(x) = x + 3$?

   a. $x - \dfrac{2}{x} + 3$

   b. $2x - \dfrac{2}{x} + 3$

   c. $x^2 - 2 + 3x - \dfrac{6}{x}$

   d. $x + 3 - \dfrac{2}{x + 3}$

4. What is the domain of $h(x) = \dfrac{x}{x^2 - 1}$?

   a. all real numbers except $x = 1$
   b. all real numbers except $x = 0$
   c. all real numbers except $x = -1$ and $x = 1$
   d. all real numbers except $x = -1$, $x = 0$, and $x = 1$

Use the following figure for questions 5 and 6.

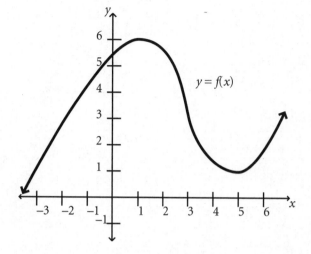

5. On what interval(s) is $f(x)$ increasing?
   a. $(-\infty, 1)$ and $(5, \infty)$
   b. $(1, 5)$
   c. $(1, 6)$
   d. $(5, \infty)$

6. Which of the following is a point of inflection for $f(x)$?
   a. $(0, 5.5)$
   b. $(1, 6)$
   c. $(3, 3)$
   d. $(5, 1)$

7. What is the equation of the straight line passing through $(2, 5)$ and $(-1, -1)$?
   a. $y = 2x + 5$
   b. $y = 2x + 1$
   c. $y = -2x + 9$
   d. $y = -2x - 3$

8. Simplify $64^{\frac{1}{2}}$.

   a. 4

   b. 8

   c. 32

   d. 4,096

9. Simplify 2^{-3}.

   a. $\dfrac{1}{8}$

   b. 8

   c. -8

   d. -6

10. Solve for x when $3^x = 15$.

   a. 5

   b. $\ln(5)$

   c. $\dfrac{\ln(15)}{\ln(3)}$

   d. $\ln(12)$

11. Evaluate $\sin\left(\dfrac{\pi}{3}\right)$.

   a. $-\dfrac{1}{2}$

   b. $\dfrac{1}{2}$

   c. $\dfrac{\sqrt{2}}{2}$

   d. $\dfrac{\sqrt{3}}{2}$

12. Evaluate $\tan\left(\dfrac{3\pi}{4}\right)$.

   a. -1

   b. 1

   c. $\dfrac{\sqrt{2}}{2}$

   d. $\sqrt{2}$

13. Evaluate $\lim\limits_{x \to 4} \dfrac{x^2 - 1}{x^2 + 1}$.

   a. -1

   b. $\dfrac{3}{5}$

   c. $\dfrac{15}{17}$

   d. $\dfrac{7}{9}$

14. Evaluate $\lim\limits_{x \to 1} \dfrac{x - 1}{x^2 - 1}$.

   a. 0

   b. 1

   c. $\dfrac{1}{2}$

   d. undefined

15. Evaluate $\lim\limits_{x \to 2^-} \dfrac{x + 3}{x - 2}$.

   a. ∞

   b. $-\infty$

   c. $-\dfrac{1}{4}$

   d. $\dfrac{-3}{2}$

16. What is the slope of $f(x) = 3x + 2$ at $x = 5$?
   a. 2
   b. 17
   c. $3x$
   d. 3

17. What is the slope of $g(x) = x^2 + 2x - 1$ at $x = 3$?
   a. 2
   b. 8
   c. 14
   d. $2x + 2$

18. Differentiate $h(x) = 4x^3 - 5x + 1$.
   a. $h'(x) = 12x^2$
   b. $h'(x) = 12x^2 - 5$
   c. $h'(x) = 12x^2 - 5x$
   d. $h'(x) = 12x^2 - 5x + \dfrac{1}{x}$

19. The height of a certain plant is $H(t) = 41 - \dfrac{40}{t}$ inches after $t \geq 1$ week. How fast is it growing after two weeks?
   a. 5 inches per week
   b. 10 inches per week
   c. 21 inches per week
   d. 31 inches per week

20. What is the derivative of $y = x^2 - 3\cos(x)$?
   a. $\dfrac{dy}{dx} = 2x + 3\sin(x)$
   b. $\dfrac{dy}{dx} = 2x - 3\sin(x)$
   c. $\dfrac{dy}{dx} = 2x - 3\cos(1)$
   d. $\dfrac{dy}{dx} = 2x - 3\tan(x)$

21. Differentiate $f(x) = \ln(x) - e^x + 2$.
   a. $f'(x) = \ln(x) + e^x$
   b. $f'(x) = \ln(x) - e^x$
   c. $f'(x) = \dfrac{1}{x} + e^x$
   d. $f'(x) = \dfrac{1}{x} - e^x$

22. Differentiate $g(x) = x^2\sin(x)$.
   a. $g'(x) = 2x\cos(x)$
   b. $g'(x) = 2x + \cos(x)$
   c. $g'(x) = 2x\sin(x) + x^2\cos(x)$
   d. $g'(x) = 2x\sin(x)\cos(x)$

23. Differentiate $j(x) = \dfrac{\ln(x)}{x}$.

    a. $j'(x) = 0$

    b. $j'(x) = \dfrac{1}{x}$

    c. $j'(x) = \dfrac{1 - \ln(x)}{x^2}$

    d. $j'(x) = \dfrac{\ln(x) - 1}{x^2}$

24. Differentiate $y = \tan(x)$.

    a. $\dfrac{dy}{dx} = \sec^2(x)$

    b. $\dfrac{dy}{dx} = -\cot(x)$

    c. $\dfrac{dy}{dx} = \dfrac{\cos^2(x) - \sin^2(x)}{\cos^2(x)}$

    d. $\dfrac{dy}{dx} = \sin(x)\cos(x)$

25. Differentiate $f(x) = e^{4x^2 + 7}$.

    a. $f'(x) = e^{8x}$

    b. $f'(x) = e^{4x^2 + 7}$

    c. $f'(x) = 8xe^{4x^2 + 7}$

    d. $f'(x) = (4x^2 + 7)e^{4x^2 - 8}$

26. Differentiate $m(x) = (x^2 - 1)^5$.

    a. $m'(x) = 10x$

    b. $m'(x) = (2x)^5$

    c. $m'(x) = 5(x^2 - 1)^4$

    d. $m'(x) = 10x(x^2 - 1)^4$

27. Compute $\dfrac{dy}{dx}$ if $y^2 + xy = x^3 + 5$.

    a. $\dfrac{dy}{dx} = x^2$

    b. $\dfrac{dy}{dx} = \dfrac{3x^2 - y}{2y + x}$

    c. $\dfrac{dy}{dx} = \dfrac{3x^2}{1 + 2y}$

    d. $\dfrac{dy}{dx} = \dfrac{3x^2 - x}{2y}$

28. Compute $\dfrac{dy}{dx}$ if $\sin(y) = 4x^2$.

    a. $\dfrac{dy}{dx} = 8x - \cos(y)$

    b. $\dfrac{dy}{dx} = 8x\cos(y)$

    c. $\dfrac{dy}{dx} = \cos(y) - 8x$

    d. $\dfrac{dy}{dx} = 8x\sec(y)$

29. What is the slope of $x^2 + y^2 = 1$ at $\left(\frac{1}{2}, \frac{\sqrt{3}}{2}\right)$?

    a. -1

    b. 1

    c. $-\dfrac{\sqrt{3}}{3}$

    d. $\dfrac{\sqrt{3}}{3}$

30. If the radius of a circle is increasing at 4 feet per second, how fast is the area increasing when the radius is 10 feet?

    a. 20π square feet per second

    b. 80π square feet per second

    c. 100π square feet per second

    d. 400π square feet per second

31. The height of a triangle increases by 3 inches every minute while its base decreases by 1 inch every minute. How fast is the area changing when the triangle has a height of 10 inches and a base of 100 inches?

    a. It is increasing at 145 square inches per minute.

    b. It is increasing at 500 square inches per minute.

    c. It is decreasing at 1,500 square inches per minute.

    d. It is decreasing at 3,000 square inches per minute.

32. Evaluate $\lim\limits_{x\to\infty} \dfrac{4x^2 - 5x + 2}{1 - x^2}$.

    a. 4

    b. -4

    c. 2

    d. undefined

33. Evaluate $\lim\limits_{x\to-\infty} \dfrac{4x^5 + 6x + 4}{x^3 + 10x - 1}$.

    a. $-\infty$

    b. ∞

    c. -4

    d. 4

34. Evaluate $\lim\limits_{x\to-\infty} \dfrac{\ln(x)}{3x + 2}$.

    a. $\dfrac{1}{3}$

    b. 2

    c. 3

    d. 0

35. Which of the following is the graph of
$$y = \frac{1}{x - 2}?$$

a.

b.

c.

d.

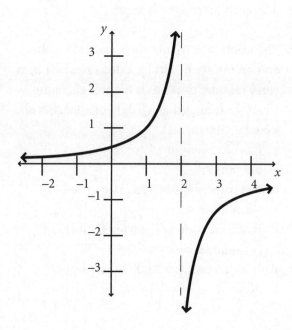

36. On what interval is $g(x) = x^4 - 6x^2 + 5$ concave down?
 a. $(1,12)$
 b. $(-6,5)$
 c. $(-\sqrt{3}, \sqrt{3})$
 d. $(-1,1)$

37. The surface area of a cube is increasing at a rate of 3 square inches per minute. How fast is an edge increasing at the instant when each side is 20 inches?
 a. $\dfrac{1}{80}$ inch per minute
 b. $\dfrac{3}{20}$ inch per minute
 c. 80 inches per minute
 d. 24,000 inches per minute

38. A box with a square bottom and no top must contain 108 cubic inches. What dimensions will minimize the surface area of the box?
 a. 2 in. × 2 in. × 27 in.
 b. 8 in. × 8 in. × 3 in.
 c. 6 in. × 6 in. × 3 in.
 d. 4 in. × 4 in. × 6.75 in.

39. If $\displaystyle\int_3^8 g(x)\,dx = 5$ and $\displaystyle\int_3^5 g(x)\,dx = -4$, then what is $\displaystyle\int_5^8 g(x)\,dx$?
 a. −20
 b. 1
 c. 3
 d. 9

40. What is $\displaystyle\int_0^4 f(x)\,dx$?

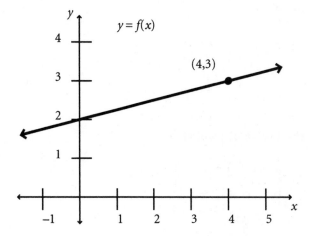

 a. 2
 b. 3
 c. 10
 d. 12

41. If $g(x)$ is the area under the curve $y = t^3 + 4t$ between $t = 0$ and $t = x$, what is $g'(x)$?
 a. $x^3 + 4x$
 b. $3x^2 + 4$
 c. $\dfrac{1}{4}x^4 + 2x$
 d. 0

42. Evaluate $\displaystyle\int (3x^2 + 8x + 5)\,dx$.
 a. $6x + 8$
 b. $6x + 8 + c$
 c. $x^3 + 4x^2 + 5x$
 d. $x^3 + 4x^2 + 5x + c$

43. Evaluate $\int_0^9 \sqrt{x}\, dx$.

   a. 3

   b. 9

   c. 18

   d. $\dfrac{81}{2}$

44. Evaluate $\int \sin(x)\, dx$.

   a. $\cos(x) + c$

   b. $-\cos(x) + c$

   c. $\sin(x) + c$

   d. $-\sin(x) + c$

45. Evaluate $\int \dfrac{x}{x^2 - 1}\, dx$.

   a. $\dfrac{\frac{1}{2}x^2}{\frac{1}{3}x^3 - x} + c$

   b. $\ln|x - 1| + c$

   c. $\dfrac{1}{2}\ln(x^2 - 1) + c$

   d. $\dfrac{1}{2}\ln|x^2 - 1| + c$

46. Evaluate $\int e^{5x}\, dx$.

   a. $\dfrac{1}{5}e^{5x} + c$

   b. $e^{5x} + c$

   c. $e^5 + c$

   d. $\dfrac{1}{5}e^5 + c$

47. Evaluate $\int 4x^2\cos(x^3)\, dx$.

   a. $4\sin(x^3) + c$

   b. $\dfrac{4}{3}\sin(x^3) + c$

   c. $\dfrac{4}{3}x^3\sin(x^3) + c$

   d. $\dfrac{4}{3}x^2\sin(x^3) + c$

48. Evaluate $\int x(x^2 + 2)^5\, dx$.

   a. $\dfrac{1}{6}(x^2 + 2)^6 + c$

   b. $5(x^2 + 2)^4 + c$

   c. $\dfrac{x^2}{2}\left(\dfrac{1}{3}x^3 + 2x\right)^6 + c$

   d. $\dfrac{1}{12}(x^2 + 5)^6 + c$

49. Evaluate $\int x\ln(x)\,dx$.

    a. $\dfrac{1}{2}x^2\ln(x) + c$

    b. $x\ln(x) - \ln(x) + c$

    c. $x^2\ln(x) + \dfrac{1}{4}x^2 + c$

    d. $\dfrac{1}{2}x^2\ln(x) - \dfrac{1}{4}x^2 + c$

50. Evaluate $\int x\sin(x)\,dx$.

    a. $-x\cos(x) + \sin(x) + c$

    b. $\dfrac{1}{2}x^2\cos(x) + c$

    c. $-\dfrac{1}{2}x^2\cos(x) + c$

    d. $x\cos(x) - \cos(x) + c$

Answers

1. **b.** Lesson 1
2. **a.** Lesson 1
3. **d.** Lesson 1
4. **c.** Lesson 1
5. **a.** Lesson 2
6. **c.** Lesson 2
7. **b.** Lesson 2
8. **b.** Lesson 3
9. **a.** Lesson 3
10. **c.** Lesson 3
11. **d.** Lesson 4
12. **a.** Lesson 4
13. **c.** Lesson 5
14. **c.** Lesson 5
15. **b.** Lesson 5
16. **d.** Lessons 6, 7
17. **b.** Lessons 6, 7
18. **b.** Lesson 7
19. **b.** Lesson 8
20. **a.** Lesson 8
21. **d.** Lesson 8
22. **c.** Lesson 9
23. **c.** Lessons 8, 9
24. **a.** Lesson 9
25. **c.** Lesson 10

26. **d.** Lesson 10
27. **b.** Lesson 11
28. **d.** Lessons 4, 11
29. **c.** Lesson 11
30. **b.** Lesson 12
31. **a.** Lesson 12
32. **b.** Lesson 13
33. **b.** Lesson 13
34. **d.** Lesson 13
35. **a.** Lesson 14
36. **d.** Lesson 14
37. **a.** Lesson 12
38. **c.** Lesson 16
39. **d.** Lesson 16
40. **c.** Lesson 16
41. **a.** Lesson 17
42. **d.** Lesson 18
43. **c.** Lesson 18
44. **b.** Lesson 18
45. **d.** Lesson 19
46. **a.** Lesson 19
47. **b.** Lesson 19
48. **d.** Lesson 19
49. **d.** Lesson 20
50. **a.** Lesson 20

1 ▶ FUNCTIONS

Calculus is the study of change. It is often important to know when a quantity is increasing, when it is decreasing, and when it hits a high or low point. Much of the business of finance depends on predicting the high and low points for prices. In science and engineering, it is often essential to know precisely how fast quantities such as temperature, size, and speed are changing. Calculus is the primary tool for calculating such changes.

Numbers, which are the focus of arithmetic, do not change. The number 5 will always be 5. It never goes up or down. Thus, we need to introduce a new sort of mathematical object, something that *can* change. These objects, the centerpiece of calculus, are functions.

Functions

A *function* is a way of matching up one set of numbers with another. The first set of numbers is called the *domain*. For each of the numbers in the domain, the function assigns exactly one number from the other set, the *range*.

It is true that in algebra, everyone is taught "parentheses mean multiplication." This means that $5(2 + 7) = 5(9) = 45$. If x is a variable, then $x(2 + 7) = x(9) = 9x$. However, if f is the name of a function, then $f(2 + 7) = f(9) =$ the number to which f takes 9. The expression $f(x)$ is pronounced "f of x" and not "f times x." This can certainly be confusing. But, as you gain experience, it will become second nature. Mathematicians use parentheses to mean several different things and expect everyone to know the difference. Sorry!

For example, the domain of the function could be the set of numbers $\{1, 4, 9, 25, 100\}$, and the range could be $\{1, 2, 3, 5, 10\}$. Suppose the function takes 1 to 1, 4 to 2, 9 to 3, 25 to 5, and 100 to 10. This could be illustrated by the following:

$$1 \rightarrow 1$$
$$4 \rightarrow 2$$
$$9 \rightarrow 3$$
$$25 \rightarrow 5$$
$$100 \rightarrow 10$$

Because we sometimes use several functions in the same discussion, it makes sense to give them names. Let us call the function we just mentioned by the name *Eugene*. Thus, we can ask, "Hey, what does Eugene do with the number 4?" The answer is "Eugene takes 4 to the number 2."

Mathematicians like to write as little as possible. Thus, instead of writing "Eugene takes 4 to the number 2," we often write "Eugene(4) = 2" to mean the same thing. Similarly, we like to use names that are as short as possible, such as f (for function), g (for function when f is already being used), h, and so on. The trigonometric functions in Lesson 4 all have three-letter names like sin and cos, but even these are abbreviations. So let us save space and use f instead of Eugene.

Because the domain is small, it is easy to write out everything:

$$f(1) = 1$$
$$f(4) = 2$$
$$f(9) = 3$$
$$f(25) = 5$$
$$f(100) = 10$$

However, if the domain were large, this would get very tedious. It is much easier to find a pattern and use that pattern to describe the function. Our function f just happens to take each number of its domain to the square root of that number. Therefore, we can describe f by saying:

f(a number) = the square root of that number

Of course, anyone with experience in algebra knows that writing "a number" over and over is a waste of time. Why not just pick a *variable* to represent the number? Just as f is a typical name for a function, little x is often used for a variable name. Using both, here is a nice way to represent our function f:

$$f(x) = \sqrt{x}$$

This tells us that putting a number into the function f is the same as putting it into $\sqrt{}$. Thus,

$$f(25) = \sqrt{25} = 5 \text{ and } f(f) = \sqrt{4} = 2.$$

Example
Find the value of $g(3)$ if $g(x) = x^2 + 2$.

Solution

Replace each occurrence of x with 3.

$$g(3) = 3^2 + 2$$

Simplify.

$$g(3) = 9 + 2 = 11$$

Example

Find the value of $h(-4)$ if $h(t) = t^3 - 2t^2 + 5$.

Solution

Replace each occurrence of t with –4.

$$h(-4) = (-4)^3 - 2(-4)^2 + 5$$

Simplify.

$$h(-4) = -64 - 2(16) + 5 = -64 - 32 + 5 = -91$$

> **⚠ CAUTION**
>
> When multiplying, an even number of negatives results in a positive number, whereas an odd number of negatives results in a negative number.

Practice

1. Find the value of $f(5)$ when $f(x) = 2x - 1$.

2. Find the value of $g(-3)$ when $g(x) = x^3 + x^2 + x + 1$.

3. Find the value of $h\left(\dfrac{1}{2}\right)$ when $h(t) = t^2 + \dfrac{3}{4}$.

4. Find the value of $f(7)$ when $f(x) = 2$.

5. Find the value of $m\left(-\dfrac{1}{5}\right)$ when $m(t) = -5t^3$.

6. Find the value of $h(64)$ when $h(x) = \sqrt{x} - \sqrt[3]{x}$.

7. Suppose that after t seconds, a rock thrown off a bridge has height $s(t) = -16t^2 + 20t + 100$ feet off the ground. What is the height above the ground after 3 seconds?

8. Suppose that the profit on making and selling x cookies is $P(x) = \dfrac{x}{2} - \dfrac{x^2}{10,000} - 10$ dollars. How much profit is made on selling 100 cookies?

Plugging Variables into Functions

Variables can be plugged into functions just as easily as numbers can. Often, though, the result can't be simplified as much.

Example

Simplify $f(w)$ if $f(x) = \sqrt{x} + 2x^2 + 2$.

Solution

Replace each occurrence of x with w.

$$f(w) = \sqrt{w} + 2w^2 + 2$$

That is all we can say without knowing more about w.

Example

Simplify $g(a + 5)$ if $g(t) = t^2 - 3t + 1$.

Solution

Replace each occurrence of t with $(a + 5)$.

$$g(a + 5) = (a + 5)^2 - 3(a + 5) + 1$$

Multiply out $(a + 5)^2$ and $-3(a + 5)$.

$$g(a + 5) = a^2 + 10a + 25 - 3a - 15 + 1$$

> **⚠ CAUTION**
>
> $(a + b)^2 \neq a^2 + b^2$. Remember to FOIL (first, outside, inside, last) to get $(a + b)^2 = a^2 + 2ab + b^2$.

Simplify.

$$g(a + 5) = a^2 + 7a + 11$$

Example

Simplify $\dfrac{f(x + a) - f(x)}{a}$ if $f(x) = x^2$.

Solution

Start with what needs to be simplified.

$$\frac{f(x + a) - f(x)}{a}$$

Use $f(x) = x^2$ to evaluate $f(x + a)$ and $f(x)$.

$$\frac{(x + a)^2 - x^2}{a}$$

Multiply out $(x + a)^2$.

$$\frac{x^2 + 2xa + a^2 - x^2}{a}$$

Cancel the x^2 and the $-x^2$.

$$\frac{2xa + a^2}{a}$$

Factor out an a.

$$\frac{(2x + a)a}{a}$$

Cancel an a from the top and bottom.

$$2x + a$$

Practice

Simplify the following.

9. $f(y)$ when $f(x) = x^2 + 3x - 1$

10. $f(x + a)$ when $f(x) = x^2 + 3x - 1$

11. $\dfrac{f(x+h) - f(x)}{h}$ when $f(x) = \dfrac{1}{2x}$

12. $g(x^2 + \sqrt{x})$ when $g(t) = \dfrac{8}{t} - 6t$

13. $g(2x) - g(x)$ when $g(t) = \dfrac{8}{t} - 6t$

14. $\dfrac{f(x+a) - f(x)}{a}$ when $f(x) = -x^2 + 5$

15. $\dfrac{h(x + a) - h(x)}{a}$ when $h(x) = -2x + 1$

16. $\dfrac{g(x+2) - g(x)}{2}$ when $g(x) = x^3$

Composition of Functions

Now that we can plug anything into functions, we can plug one function in as the input of another function. This is called *composition*. The composition of function f with function g is written $f \circ g$. This means to plug g into f like this:

$$(f \circ g)(x) = f(g(x))$$

It may seem that f comes first in $(f \circ g)(x)$, reading from left to right, but actually, the g is closer to the x. This means that the function g acts on the x first.

Example

If $f(x) = \sqrt{x} + 2x$ and $g(x) = 4x + 7$, then what is the composition $(f \circ g)(x)$?

Solution

Start with the definition of composition.

$$(f \circ g)(x) = f(g(x))$$

Use $g(x) = 4x + 7$.

$$(f \circ g)(x) = f(4x + 7)$$

Replace each occurrence of x in f with $4x + 7$.

$$(f \circ g)(x) = \sqrt{4x + 7} + 2(4x + 7)$$

Simplify.

$$(f \circ g)(x) = \sqrt{4x + 7} + 8x + 14$$

Conversely, to evaluate $(g \circ f)(x)$, we compute:

$$(g \circ f)(x) = g(f(x))$$

Use $f(x) = \sqrt{x} + 2x$.

$$(g \circ f)(x) = g(\sqrt{x} + 2x)$$

Replace each occurrence of x in g with $\sqrt{x} + 2x$.

$$(g \circ f)(x) = 4(\sqrt{x} + 2x) + 7$$

Simplify.

$$(g \circ f)(x) = 4\sqrt{x} + 8x + 7$$

⚠ CAUTION

This shows that the order in which you compute a composition matters! In general, $(f \circ g)(x) \neq (g \circ f)(x)$.

We can form the composition of more than two functions. Just apply the functions, one at a time, working your way from the one closest to x outward.

Example
If $f(x) = \dfrac{x+1}{2x-3}$, $g(x) = 2 - x$, and $h(x) = 4x$, then what is $(f \circ g \circ h)(x)$?

Solution
Start with the definition of composition.

$$(f \circ g \circ h)(x) = f(g(h(x)))$$
Use $h(x) = 4x$.

$$(f \circ g \circ h)(x) = f(g(4x))$$
Compute $g(4x)$ by replacing each occurrence of x in g with $4x$.

$$g(4x) = 2 - 4x$$

Next, substitute this into the composition.

$$(f \circ g \circ h)(x) = f(g(4x)) = f(2 - 4x).$$
Replace every occurrence of x in f with $2 - 4x$.

$$(f \circ g \circ h)(x) = f(2 - 4x) = \frac{(2-4x)+1}{2(2-4x)-3}$$
Simplify.

$$(f \circ g \circ h)(x) = \frac{3-4x}{1-8x}$$

Practice

Using $f(x) = \dfrac{1}{x}$, $g(x) = x^3 - 2x^2 + 1$, and $h(x) = x - \sqrt{x}$, simplify the following compositions.

17. $(f \circ g)(x)$

18. $(g \circ f)(x)$

19. $(f \circ h)(t)$

20. $(f \circ f)(z)$

21. $(h \circ h)(w)$

22. $(g \circ h)(16)$

23. $(h \circ f \circ g)(x)$

24. $(f \circ h \circ f)(2x)$

Domains

When an expression is used to describe a function $f(x)$, it is convenient to think of the domain as the set of all numbers that can be substituted into the expression and get a meaningful output. This set is called the *domain*. The range of the function is the set of all possible numbers produced by evaluating f at the numbers in its domain.

In the beginning of the lesson, we considered the function:

$$f(x) = \sqrt{x}$$

However, we left out a crucial piece of information: the domain. The domain of this function consisted of only the numbers 1, 4, 9, 25, and 100. Thus, we should have written

$$f(x) = \sqrt{x} \text{ if } x = 1, 4, 9, 25, \text{ or } 100$$

Usually, the domain of a function is not given explicitly like this. In such situations, it is assumed that the domain is as large as it possibly can be, meaning that it contains all real numbers that, when plugged into the function, produce another real number. Specifically, including a number in the domain cannot violate one of the following two fundamental prohibitions:

- Never divide by zero.
- Never take an even root of a negative number.

Example

What is the domain of $f(x) = \dfrac{3}{x-2}$?

Solution

We must never let the denominator $x - 2$ be zero, so x cannot be 2. Therefore, the domain of this function consists of all real numbers except 2.

The prohibition against even roots (like square roots) of negative numbers is less severe. An even root of a negative number is an imaginary number. Useful mathematics can be done with imaginary numbers. However, for the sake of simplicity, we will avoid them in this book.

Example

What is the domain of $g(x) = \sqrt{3x + 2}$?

Solution

The numbers in the square root must not be negative, so $3x + 2 \geq 0$, thus $x \geq -\dfrac{2}{3}$. The domain consists of all numbers greater than or equal to $-\dfrac{2}{3}$.

Do note that it is perfectly okay to take the square root of zero, since $\sqrt{0} = 0$. It is only when numbers are less than zero that even roots become imaginary.

Example

Find the domain of $k(x) = \dfrac{\sqrt{4 - x}}{x^2 + 5x + 6}$.

Solution

To avoid dividing by zero, we need $x^2 + 5x + 6 \neq 0$, so $(x + 3)(x + 2) \neq 0$, thus $x \neq -3$ and $x \neq -2$. To avoid an even root of a negative number, $4 - x \geq 0$, so $x \leq 4$. Thus, the domain of k is $x \leq 4$, $x \neq -3$, $x \neq -2$.

A nice way of representing certain collections of real numbers is interval notation, as follows:

COLLECTION OF REAL NUMBERS	INTERVAL NOTATION
$a < x < b$	(a,b)
$a \leq x < b$	$[a,b)$
$a < x \leq b$	$(a,b]$
$a \leq x \leq b$	$[a,b]$
$x > a$	(a,∞)
$x \geq a$	$[a,\infty)$
$x < b$	$(-\infty,b)$
$x \leq b$	$(-\infty,b]$
All real numbers	$(-\infty,\infty)$

Note: A parenthesis is used when we intend to NOT include a point, whereas a square bracket is used when we intend TO include a point.

The domain of the previous example would be written as follows:

$(-\infty,-3)$, $(-3,-2)$, and $(-2,4]$

Practice

Find the domain of each of the following functions. Express your answers using interval notation.

25. $f(x) = \dfrac{-1}{(x+3)(x-5)^2}$

26. $h(x) = \sqrt{x + 1}$

27. $k(t) = \dfrac{1}{\sqrt{t+5}}$

28. $g(x) = x^2 + 5x - 6$

29. $j(z) = \dfrac{(z-1)(z+2)}{z^2+1}$

30. $h(x) = \sqrt[3]{x}$

31. $k(x) = \dfrac{\sqrt[4]{2-x}}{x+8}$

32. $g(u) = \dfrac{8u}{(u+3)\sqrt{4+3u}}$

GRAPHS

LESSON 2

A function can be fully described by showing explicitly what happens at each number in its domain (for example, $4 \rightarrow 2$) or by giving its formula (for example, $f(x) = \sqrt{x}$). However, neither of these provides a clear visual picture of the function.

Fortunately, René Descartes came up with the idea of a *graph*, a visual picture of a function. Rather than say $4 \rightarrow 2$ or $f(4) = 2$, we plot the point $(4,2)$ on the Cartesian plane, as in Figure 2.1.

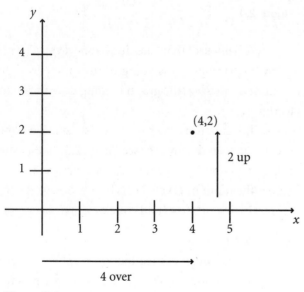

Figure 2.1

We put the y into the formula $y = f(x) = \sqrt{x}$ to imply that the y-coordinates of our points are the numbers we get by plugging the x-coordinates into the function f.

Practice

Plot the following points on a Cartesian plane.

1. $(3,5)$

2. $(-3,4)$

3. $(2,-6)$

4. $(-1,-5)$

5. $(0,3)$

6. $(-5,0)$

7. $(0,0)$

8. $\left(\dfrac{9}{2}, \dfrac{1}{4}\right)$

For the function $f(x) = x^2 - 2x + 5$, plot the point $(x, f(x))$ for the following values of x.

9. $x = 3$

10. $x = 1$

11. $x = 0$

12. $x = -2$

If we plotted the points $(x, f(x))$ for all x in the domain of $f(x) = \sqrt{x}$ (not just the whole numbers, but all the fractions and decimals, too), then the points would be so close together that they would form a continuous curve as in Figure 2.2.

The graph shows us several interesting characteristics of the function $f(x) = \sqrt{x}$.

Figure 2.2

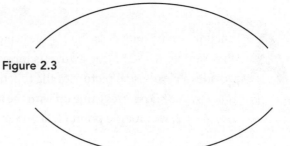

Figure 2.3

Figure 2.4

We can see that the function $f(x) = \sqrt{x}$ is *increasing* (the graph is going up from left to right) and not *decreasing* (the graph is going down from left to right).

The function $f(x) = \sqrt{x}$ is *concave down* because it bows downward (see Figure 2.3) like a frown and not *concave up* like a smile (see Figure 2.4). We report the input intervals in each case. So, we say that f is increasing on $(0,\infty)$ and concave down on $(0,\infty)$.

Example

Assume the domain of the function graphed in Figure 2.5 is all real numbers. Determine where the function is increasing and decreasing, and where the function is concave up and concave down.

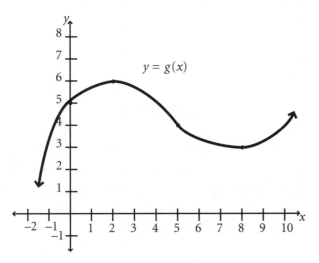

Figure 2.5

There is no way to tell the difference without an actual formula for $f(x)$.

The point at (2,6) where g stops increasing and begins to decrease is the highest point in its immediate vicinity and is called a *local maximum*. The point at (8,3) is similarly a *local minimum*, the lowest point in its neighborhood. These points tend to be of particular interest, especially in applications.

Example

Use the graph of the function $h(x)$ in Figure 2.6 to identify the domain, where it is increasing and decreasing, where it has local maxima and minima, where it is concave up and down, and where it has points of inflection.

Solution

The function g is increasing up to the point at $x = 2$, where it then decreases down to $x = 8$, and then increases thereafter. Using interval notation, we say that g increases on $(-\infty,2)$ and on $(8,\infty)$, and that g decreases on (2,8).

The concavity of g is trickier to estimate. Clearly g is concave down in the vicinity of $x = 2$ and concave up starting around $x = 7$. The exact point where the concavity changes is called a *point of inflection*. On this graph, it seems to be at the point (5,4), though some people might imagine it to actually be a bit on either side. Thus, we say that g is concave down on $(-\infty,5)$ and concave up on $(5,\infty)$.

Honestly, any information obtained by simply eyeballing a graph is going to be a rough estimate. Is the local maximum at (2,6), or is it at (2.0003,5.9998)?

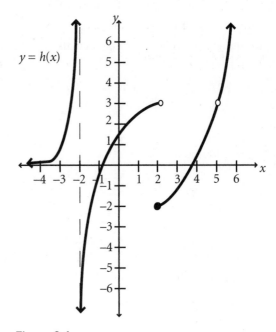

Figure 2.6

Solution

The first thing to notice is that *h* has three breaks, or *discontinuities*. If we wanted to trace the graph of *h* with a continuous motion of a pencil, then we would have to lift up the pencil at $x = -2$, $x = 2$, and at $x = 5$. The little unshaded circle at (5,3) indicates a hole in the graph where a single point has been taken out. This means that $x = 5$ is not in the domain, just as $x = -2$ has no point above or below it. The situation at $x = 2$ is more interesting because $x = 2$ *is* in the domain, with the point (the shaded-in circle) at $(2, -2)$ representing $h(2) = -2$. All of the points immediately before $x = 2$ have *y*-values close to $y = 3$, but then there is an abrupt jump down to $x = 2$. Such jumps occur often when describing real-life situations using functions like the way the cost of postage leaps up as soon as a letter weighs more than one ounce.

Because of the discontinuities, we must name each interval separately, as in: *h* increases on $(-\infty, -2)$, $(-2, 2)$, $(2, 5)$, and on $(5, \infty)$. As well, *h* is concave up on $(-\infty, -2)$, $(2, 5)$, and on $(5, \infty)$, and concave down on $(-2, 2)$.

There is a local minimum at $(2, -2)$, because this point there is the lowest in its immediate vicinity, say for all $1 < x < 3$. There is no local maximum in that range because the *y*-values get really close to $y = 3$; there is no highest point in the range because of the unshaded circle.

Similarly, a point of inflection can be seen at $x = 2$ but not at $x = -2$ because there can't be a point of inflection where there is no point!

The line $x = -2$ is called a *vertical asymptote* because the graph of $f(x)$ begins to look more like this line the closer the inputs get to -2. Because the graph appears to flatten out like the straight horizontal line $y = 0$ (the *x*-axis) as the graph goes off to the left, we say that the graph of $y = h(x)$ appears to have a *horizontal asymptote* at $y = 0$. We will examine this more closely in Lesson 13.

Practice

Use the graph of each function to determine the discontinuities, where the function is increasing and decreasing, the local maximum and minimum points, where the function is concave up and down, the points of inflection, and the asymptotes.

13.

14.

15.

16.

17.

18.

19.

20.

Note

We can obtain all sorts of useful information from a graph, such as its maximal points, where it is increasing and decreasing, and so on. Calculus will enable us to get this information directly from the function. We will then be able to draw graphs intelligently, without having to calculate and plot thousands of points (the method graphing calculators use).

Straight Lines

The most familiar and arguably most widely used of all graphs are straight lines. Human beings tend to build and move linearly. Given any two points, we can immediately get a feel for the steepness of a line, as seen in Figure 2.7.

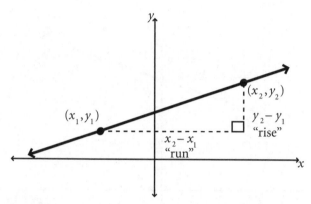

Figure 2.7

"How much a line is increasing or decreasing" is called the *slope* and is calculated by:

$$slope = \frac{rise}{run}$$

$$= \frac{y\text{-}change}{x\text{-}change}$$

$$= \frac{y_2 - y_1}{x_2 - x_1}$$

⚠ CAUTION

Make sure to subtract the *y*'s in the top and the *x*'s in the bottom IN THE SAME ORDER. For instance, don't use $y_2 - y_1$ for the rise and $x_1 - x_2$ for the run.

Example

What is the slope of the line passing through points $(2,7)$ and $(-1,5)$?

Solution

$$slope = \frac{5 - 7}{-1 - 2} = \frac{-2}{-3} = \frac{2}{3}$$

Practice

Find the slope between the following points.

21. $(1,5)$ and $(2,8)$

22. $(7,3)$ and $(-2,3)$

23. $(-2,-4)$ and $(-6,5)$

24. $(2,7)$ and $(5,w)$

Equation of a Line

No matter what two points you choose on a line, the slope will always be the same. Thus, if a straight line has slope m and goes through the point (x_1, y_1), then using any other point (x, y) on the line, we get the same slope, namely:

$$\frac{y - y_1}{x - x_1} = m$$

By cross-multiplying, we get the *point-slope formula* for the equation of a straight line:

$$y - y_1 = m(x - x_1)$$
or equivalently
$$y = m(x - x_1) + y_1$$

Going one step further, we get

$$y = mx + \underbrace{(-mx_1 + y_1)}_{\text{Call this } b}$$

This is called the *slope-intercept formula* for the line because the point $(0, b)$ is the y-intercept of the line (that is, where it crosses the y-axis).

Example

Find the equation of the line with slope -2 through point $(-1,8)$. Graph the line.

Solution

$y = -2(x - (-1)) + 8$

$y = -2x + 6$ (see Figure 2.8)

Figure 2.8

The slope of $-2 = \dfrac{-2}{1}$ means the y-value goes down 2 with every decrease of 1 unit in the x-value.

Example

Find the equation of the straight line through $(2,6)$ and $(5,7)$. Graph the line.

Solution

The slope is $\dfrac{7 - 6}{5 - 2} = \dfrac{1}{3}$, so the equation is $y = \dfrac{1}{3}(x - 2) + 6 = \dfrac{1}{3}x + \dfrac{16}{3}$ (see Figure 2.9).

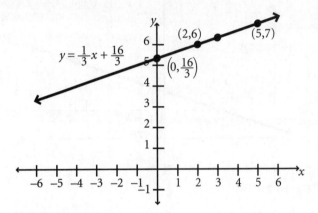

Figure 2.9

The slope of $\dfrac{1}{3}$ means the y-value goes up 1 for every increase of 3 units in the x-value.

Practice

Find the equation of the straight line with the given information and then graph the line.

25. slope 2 through point $(1,-2)$

26. slope $-\dfrac{2}{3}$ through point $(6,1)$

27. through points $(5,3)$ and $(-1,-3)$

28. through points $(2,5)$ and $(6,5)$

3 ▶ EXPONENTS AND LOGARITHMS

Exponents

Exponents frequently arise in calculations throughout calculus. If a is a positive real number and n is a positive integer (that is, $n = 1, 2, 3, \ldots$), then a^n means "multiply the base a by itself n times." Symbolically,

$$a^n = \underbrace{a \cdot a \cdot a \cdots a}_{n \text{ times}}$$

⚠ CAUTION

Do not multiply the *base a* times the *exponent n*.
Symbolically, $a^n \neq a \cdot n$.

Examples

Review the following examples.

$$3^4 = 3 \cdot 3 \cdot 3 \cdot 3 = 81$$

$$2^5 = 2 \cdot 2 \cdot 2 \cdot 2 \cdot 2 = 32$$

$5^1 = 5$

$10^6 = 1,000,000$

$\left(\dfrac{1}{2}\right)^3 + \dfrac{1}{2} \cdot \dfrac{1}{2} \cdot \dfrac{1}{2} = \dfrac{1}{8}$

When two numbers with the same base are multiplied, their exponents are added.

$$a^n \cdot a^m = (\underbrace{a \cdot a \cdot a \cdots a}_{n \text{ times}}) \cdot (\underbrace{a \cdot a \cdot a \cdots a}_{m \text{ times}}) = a^{n+m}$$

Examples
Review the following examples.

$4^{10} \cdot 4^7 = 4^{17}$

$\left(\dfrac{2}{3}\right)^2 \cdot \left(\dfrac{2}{3}\right)^2 = \dfrac{2}{3} \cdot \dfrac{2}{3} \cdot \dfrac{2}{3} \cdot \dfrac{2}{3} = \dfrac{16}{81}$

$5^3 \cdot 5 = 5^3 \cdot 5^1 = 5^4$

$7^2 \cdot 7^4 \cdot 7^3 = 7^9$

The rule about adding exponents has an interesting consequence. We know that $\sqrt{5} \cdot \sqrt{5} = 5$ because this is what "square root" means. Alternately, $5^{\frac{1}{2}} \cdot 5^{\frac{1}{2}} = 5^{\frac{1}{2}+\frac{1}{2}} = 5^1 = 5$. Because $\sqrt{5}$ and $5^{\frac{1}{2}}$ act exactly the same, they are equal: $\sqrt{5} = 5^{\frac{1}{2}}$. This works for square roots, cube roots, and so on:

$$a^{\frac{1}{2}} = \sqrt{a}, \ a^{\frac{1}{3}} = \sqrt[3]{a}, \ a^{\frac{1}{4}} = \sqrt[4]{a}, \ \ldots$$

Examples
Review the following examples.

$9^{\frac{1}{2}} = \sqrt{9} = 3$

$64^{\frac{1}{3}} = \sqrt[3]{64} = 4$

When two numbers with the same base are divided, their exponents are subtracted.

$$\dfrac{a^n}{a^m} = a^{n-m}$$

Examples
Work through the following simplifications.

$$\dfrac{3^5}{3^2} = \dfrac{3 \cdot 3 \cdot 3 \cdot 3 \cdot 3}{3 \cdot 3} = \dfrac{3 \cdot 3 \cdot 3 \cdot \cancel{3} \cdot \cancel{3}}{\cancel{3} \cdot \cancel{3}} = 3 \cdot 3 \cdot 3 = 3^3$$

$$\dfrac{11^{15}}{11^6} = 11^{15-6} = 11^9$$

The rule about subtracting exponents has two interesting consequences. First, $\dfrac{5^4}{5^4} = 1$ because any nonzero number divided by itself is one. Also, $\dfrac{5^4}{5^4} = 5^{4-4} = 5^0$. Thus, $5^0 = 1$. In general:

$$a^0 = 1$$

> **⚠ CAUTION**
> Again, don't multiply a times the exponent 0 to conclude that $a^0 = 0$.

Simplify the following.

$3^0 = 1$

$200^0 = 1$

The second consequence follows from:

$$\dfrac{2^3}{2^7} = \dfrac{2 \cdot 2 \cdot 2}{2 \cdot 2 \cdot 2 \cdot 2 \cdot \cancel{2} \cdot \cancel{2} \cdot \cancel{2}} = \dfrac{1}{2 \cdot 2 \cdot 2 \cdot 2} = \dfrac{1}{2^4} \text{ while}$$

also $\dfrac{2^3}{2^7} = 2^{3-7} = 2^{-4}$. Thus, $2^{-4} = \dfrac{1}{2^4}$. In general:

$$a^{-n} = \dfrac{1}{a^n}$$

Examples
Review the following examples.

$3^{-2} = \dfrac{1}{3^2} = \dfrac{1}{9}$

$4^{-1} = \dfrac{1}{4^1} = \dfrac{1}{4}$

$5^{-\frac{1}{2}} = \dfrac{1}{5^{\frac{1}{2}}} = \dfrac{1}{\sqrt{5}}$

When a^n is itself raised to a power m, the exponents are multiplied.

$$\left(a^n\right)^m = \underbrace{a^n \cdot a^n \cdot \ldots \cdot a^n}_{m \text{ times}} = a^{\overbrace{n+\ldots+n}^{m \text{ times}}} = a^{n \cdot m}$$

Examples

Review the following examples.

$$\left(5^2\right)^2 = 5^2 \cdot 5^2 = 5^{2+2} = 5^4$$

$$\left(4^{-2}\right)^3 = 4^{-2} \cdot 4^{-2} \cdot 4^{-2} = 4^{(-2)+(-2)+(-2)} = 4^{-6} = \frac{1}{4^6}$$

Practice

Simplify the following.

1. $2^3 \cdot 2^2$

2. $4 \cdot 4^2$

3. $\dfrac{10^7}{10^3}$

4. $\dfrac{6^3}{6^5}$

5. 6^0

6. $3^8 \cdot 3 \cdot 3^{-5}$

7. 9^1

8. $5{-}1 \cdot 5$

9. 2^{-3}

10. $27^{\frac{2}{3}}$

11. $\left(5^{-2}\right)^3$

12. $\dfrac{\left(8^{-4}\right)^{\frac{1}{2}}}{8^2}$

13. $\dfrac{4^{-2} \cdot 4^0 \cdot 4^5}{4^{-1}}$

14. $\dfrac{\left(9^{\frac{1}{2}}\right)^{-2}}{81^{-2}}$

Exponential Functions

We can form an *exponential function* by leaving the base fixed and varying the exponent.

Example

The function $f(x) = 2^x$ has the graph shown in Figure 3.1. Note that 2^x is quite different from x^2. For example, when $x = 10$, the value of 2^x is $2^{10} = 2 \cdot 2 \cdot 2 \cdot 2 \cdot 2 \cdot 2 \cdot 2 \cdot 2 \cdot 2 \cdot 2 = 1{,}024$, while the value of x^2 is $10^2 = 10 \cdot 10 = 100$.

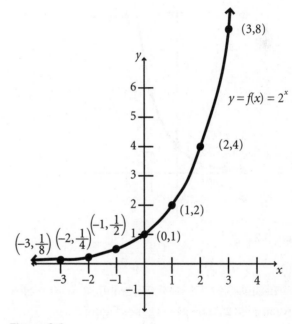

Figure 3.1

Example

The function $g(x) = 3^x$ has the graph shown in Figure 3.2. Note that $g(x)$ grows faster than $f(x) = 2^x$ as x gets larger. For reasons that will become clear later, a very nice base to use is the number $e = 2.71828\ldots$, which, just like $\pi = 3.14159\ldots$, can never be written out completely.

The exponential function $f(x) = e^x$ and the natural logarithm $g(x) = \ln(x)$ "undo" each other when composed. That is,

$f(g(x)) = e^{\ln(x)} = x$ and $g(f(x)) = \ln(e^x) = x$

We say that f and g are *inverses*.

Figure 3.2

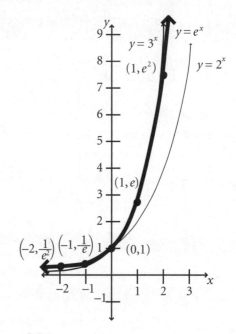

Figure 3.3

Because $2 < e < 3$, the graph of $y = e^x$ fits between $y = 2^x$ and $y = 3^x$ (see Figure 3.3).

Another useful function is the inverse of e^x, known as the *natural logarithm* $\ln(x)$. Just as subtracting undoes adding, dividing undoes multiplying, and taking a square root undoes squaring, the natural logarithm undoes e^x.

If $y = e^x$, then $\ln(y) = \ln(e^x) = x$.

The graph of $y = \ln(x)$ comes from flipping the graph of $y = e^x$ across the line $y = x$, as depicted in Figure 3.4. In particular, since $e^0 = 1$, it follows that $\ln(1) = 0$.

⚠ CAUTION

$\ln(0) \neq 1$! In fact, ln is not even defined at $x = 0$.

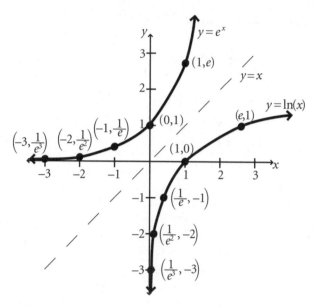

Figure 3.4

The laws of natural logarithms might appear unusual, but they are natural consequences of the exponent rules.

$$\ln(a) + \ln(b) = \ln(a \cdot b)$$

$$\ln(a) - \ln(b) = \ln\left(\frac{a}{b}\right)$$

$$\ln(a^n) = n \cdot \ln(a)$$

> ⚠ **CAUTION**
>
> These are the only three properties. Take particular note of the following, which are often mistakenly used in their place.
> $\ln(a + b) \neq \ln(a) + \ln(b)$
> $\ln(a - b) \neq \ln(a) - \ln(b)$
> $(\ln(a))^b \neq b \ln(a)$

The last of the three preceding laws is useful for turning an exponent into a matter of multiplication.

Example
Solve for x when $10^x = 7$.

Solution
Take the natural logarithm of both sides.

$$\ln(10^x) = \ln(7)$$

Use $\ln(a^n) = n \cdot \ln(a)$.

$$x \cdot \ln(10) = \ln(7)$$

Divide both sides by $\ln(10)$.

$$x = \frac{\ln(7)}{\ln(10)}$$

A calculator can be used to find a decimal approximation: $\frac{\ln(7)}{\ln(10)} \approx 0.84509$, if desired.

Example
Simplify $\ln(25) + \ln(4) - \ln(2)$.

Solution
Use $\ln(a) + \ln(b) = \ln(a \cdot b)$.

$$\ln(25 \cdot 4) - \ln(2)$$

Use $\ln(a) - \ln(b) = \ln\left(\frac{a}{b}\right)$.

$$\ln\left(\frac{25 \cdot 4}{2}\right) = \ln(50)$$

Practice

Simplify the following.

15. $e^3 \cdot e^8$

16. $\dfrac{e^{12}}{e^5}$

17. e^0

18. $\ln(e^2)$

19. $e^{\ln(5)}$

20. $\ln\left(25^0\right)$

21. $\ln(7) \, 1 \ln(2)$

22. $\ln(24) \, 2 \ln(6)$

23. $\ln(5) - \ln(2) + \ln\left(\dfrac{2}{5}\right)$

24. Solve for x when $2^x = 10$.

25. Solve for x when $3^x \cdot 3^5 = 100$.

26. If $\log_a x = 2$ and $\log_a y = -3$, then what is

$\log_a\left(\dfrac{x}{y^3}\right)^?$

4 ▶ TRIGONOMETRY

Some very interesting and important functions are formed by dividing the length of one side of a right triangle by the length of another side. These functions are called *trigonometric* because they come from the geometry of a right triangle. Let *H* represent the length of the *hypotenuse*, *A* represent the length of the side *adjacent* to the angle *x*, and *O* represent the length of the side *opposite* (away) from the angle *x*. Such a triangle is depicted in Figure 4.1.

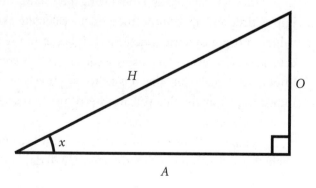

Figure 4.1

Some people remember the first three trigonometric functions by *SOA CAH TOA* to remember $\sin(x) = \dfrac{O}{A}$, $\cos(x) = \dfrac{A}{H}$, and $\tan(x) = \dfrac{O}{A}$.

The six trigonometric functions, *sine* (abbreviated sin), *cosine* (cos), *tangent* (tan), *secant* (sec), *cosecant* (csc), and *cotangent* (cot), are defined at an angle *x* by dividing the following sides:

$$\sin(x) = \frac{O}{H}$$

$$\cos(x) = \frac{A}{H}$$

$$\tan(x) = \frac{O}{A}$$

$$\sec(x) = \frac{H}{A}$$

$$\csc(x) = \frac{H}{O}$$

$$\cot(x) = \frac{A}{O}$$

The first thing to notice is that all of the functions can be obtained from just $\sin(x)$ and $\cos(x)$ using the following *trigonometric identities.*

$$\tan(x) = \frac{O}{A} = \frac{\frac{O}{H}}{\frac{A}{H}} = \frac{\sin(x)}{\cos(x)}$$

$$\sec(x) = \frac{H}{A} = \frac{1}{\frac{A}{H}} = \frac{1}{\cos(x)}$$

$$\csc(x) = \frac{H}{O} = \frac{1}{\frac{O}{H}} = \frac{1}{\sin(x)}$$

$$\cot(x) = \frac{A}{O} = \frac{\frac{A}{H}}{\frac{O}{H}} = \frac{\cos(x)}{\sin(x)}$$

Thus, all of the trigonometric functions can be evaluated for an angle *x* if the $\sin(x)$ and $\cos(x)$ are known.

The next thing to notice is that the Pythagorean theorem, which, stated in terms of the sides *O*, *A*, and *H*, is $O^2 + A^2 = H^2$. And, if we divide through by H^2, we get the following:

$$\frac{O^2}{H^2} + \frac{A^2}{H^2} = \frac{H^2}{H^2}$$

$$\left(\frac{O}{H}\right)^2 + \left(\frac{A}{H}\right)^2 = 1$$

Thus, $(\sin(x))^2 + (\cos(x))^2 = 1$. To save on parentheses, we often write this as $\sin^2(x) + \cos^2(x) = 1$. Because no particular value of *x* was used in the calculations, this is true for every value of *x*.

Drawing triangles and measuring their sides is an impractical and inaccurate method to calculate the values of trigonometric functions. It is better to use a calculator. However, when using a calculator, it is very important to make sure that it is set to the same format for measuring angles that you are already using: that is, *degrees* or *radians.*

There are 360 degrees in a circle, possibly because ancient peoples thought that there were 360 days in a

To convert from degrees to radians, multiply by $\dfrac{2\pi}{360°} = \dfrac{\pi}{180°}$.

To convert from radians to degrees, multiply by $\dfrac{360°}{2\pi} = \dfrac{180°}{\pi}$.

year. As the earth went around the sun, the position of the sun against the background stars moved one *degree* every day. The 2π radians in a circle correspond to the distance around a circle of radius 1.

- To convert from degrees to radians, multiply by
  $$\frac{2\pi}{360°} = \frac{\pi}{180°}.$$
- To convert from radians to degrees, multiply by
  $$\frac{360°}{2\pi} = \frac{180°}{\pi}.$$

Example
Convert 45° to radians.

Solution
$$45° = 45° \cdot \frac{\pi}{180°} \text{ radians} = \frac{\pi}{4} \text{ radians}$$

Example
Convert $\dfrac{2\pi}{3}$ radians to degrees.

Solution
$$\frac{2\pi}{3} \text{ radians} = \frac{2\pi}{3} \cdot \frac{180°}{\pi} = 120°$$

Practice

Convert the following to radians.

1. 30°

2. 180°

3. 270°

4. 300°

5. 135°

Convert the following to degrees.

6. $\dfrac{\pi}{3}$

7. $\dfrac{\pi}{2}$

8. 2π

9. $\dfrac{\pi}{10}$

10. $\dfrac{11\pi}{6}$

Trigonometric Values of Nice Angles

There are a few nice angles for which the trigonometric functions can be easily calculated. If $x = \dfrac{\pi}{4} = 45°$, then the two legs of the triangle are equal. If the hypotenuse is $H = 1$, then we have the triangle in Figure 4.2.

Another nice angle is $x = 60° = \dfrac{\pi}{3}$, because it is found in equilateral triangles such as the one seen in Figure 4.3. This triangle can be cut in half by inserting a segment from the top vertex down to the base, resulting in the triangle shown in Figure 4.4.

Figure 4.3

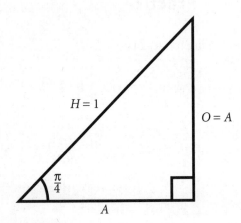

Figure 4.2

By the Pythagorean theorem, $A^2 + A^2 = 1$, so $2A^2 = 1$ and $A^2 = \dfrac{1}{2}$. This means that $O = A = \dfrac{1}{\sqrt{2}} = \dfrac{1}{\sqrt{2}}$. If we rationalize the denominator, we get $\dfrac{1}{\sqrt{2}} = \dfrac{1}{\sqrt{2}} \cdot \dfrac{\sqrt{2}}{\sqrt{2}} = \dfrac{\sqrt{2}}{2}$. Thus:

$$\sin\left(\dfrac{\pi}{4}\right) = \dfrac{O}{H} = \dfrac{\frac{\sqrt{2}}{2}}{1} = \dfrac{\sqrt{2}}{2}$$

$$\cos\left(\dfrac{\pi}{4}\right) = \dfrac{A}{H} = \dfrac{\frac{\sqrt{2}}{2}}{1} = \dfrac{\sqrt{2}}{2}$$

Figure 4.4

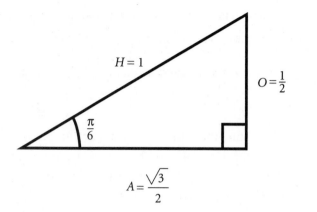

Figure 4.5

By the Pythagorean theorem, $\left(\frac{1}{2}\right)^2 + O^2 = 1^2$,

so $O^2 = 1 - \frac{1}{4} = \frac{3}{4}$. Thus, $O = \sqrt{\frac{3}{4}} = \frac{\sqrt{3}}{2}$. This

means that:

$$\sin\left(\frac{\pi}{3}\right) = \frac{O}{H} = \frac{\frac{\sqrt{3}}{2}}{1} = \frac{\sqrt{3}}{2}$$

$$\cos\left(\frac{\pi}{3}\right) = \frac{A}{H} = \frac{\frac{1}{2}}{1} = \frac{1}{2}$$

We can flip that last triangle around to calculate the trigonometric functions for the other angle $x = 30°$ $= \frac{\pi}{6}$ (see Figure 4.5).

$$\sin\left(\frac{\pi}{6}\right) = \frac{O}{H} = \frac{\frac{1}{2}}{1} = \frac{1}{2}$$

$$\cos\left(\frac{\pi}{6}\right) = \frac{A}{H} = \frac{\frac{\sqrt{3}}{2}}{1} = \frac{\sqrt{3}}{2}$$

Example
Compute $\sec\left(\frac{\pi}{4}\right)$.

Solution
Use the trigonometric identity for sec.

$$\sec(x) = \frac{1}{\cos(x)}$$

Use $x = \frac{\pi}{4}$.

$$\sec\left(\frac{\pi}{4}\right) = \frac{1}{\cos(\frac{\pi}{4})}$$

Use $\cos\left(\frac{\pi}{4}\right) = \frac{\sqrt{2}}{2}$.

$$\sec\left(\frac{\pi}{4}\right) = \frac{1}{\frac{\sqrt{2}}{2}}$$

Simplify.

$$\sec\left(\frac{\pi}{4}\right) = \frac{2}{\sqrt{2}} = \frac{2}{\sqrt{2}} \cdot \frac{\sqrt{2}}{\sqrt{2}} = \frac{2\sqrt{2}}{2} = \sqrt{2}$$

Practice

Use the trigonometric identities to evaluate the following.

11. $\tan\left(\dfrac{\pi}{4}\right)$

12. $\tan\left(\dfrac{\pi}{3}\right)$

13. $\csc\left(\dfrac{\pi}{6}\right)$

14. $\sec\left(\dfrac{\pi}{3}\right)$

15. $\cot\left(\dfrac{\pi}{3}\right)$

16. $\cot\left(\dfrac{\pi}{6}\right)$

17. $\sec\left(\dfrac{\pi}{6}\right)$

18. $\csc\left(\dfrac{\pi}{4}\right)$

Trigonometric Values for Angles Greater Than 90° = $\dfrac{\pi}{2}$

Example

For example, for $\dfrac{\pi}{6} = 30°$, we have the picture shown in Figure 4.6.

$$\sin\left(\frac{\pi}{6}\right) = \frac{1}{2}$$

$$\cos\left(\frac{\pi}{6}\right) = \frac{\sqrt{3}}{2}$$

The circle of radius 1 around the origin is called the *unit circle*. As such, the hypotenuse has length 1, and the sine is the *y*-value of the point where a ray of the given angle intersects with the circle of radius 1. Similarly, the cosine is the *x*-value. Note in Figure 4.7 that the angle of measure 0 runs straight to the right along the positive *x*-axis, and every other positive angle is measured counterclockwise from there.

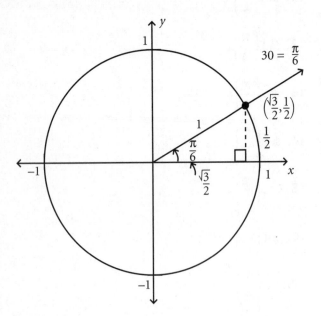

Figure 4.6

This can be used to find the trigonometric values of nice angles greater than 90° = $\dfrac{\pi}{2}$. The trick is to use either a 30°, 60°, 90° triangle (a $\dfrac{\pi}{6}$, $\dfrac{\pi}{3}$, $\dfrac{\pi}{2}$ triangle) or else a 45°, 45°, 90° triangle (a $\dfrac{\pi}{4}$, $\dfrac{\pi}{4}$, $\dfrac{\pi}{2}$ triangle) to find the *y*-value(sine) and *x*-value (cosine) of the appropriate point on the unit circle. As before, calculating the trigonometric values for non-nice angles is best done with a calculator.

Example

Find the sine and cosine of 120° = $\dfrac{2\pi}{3}$.

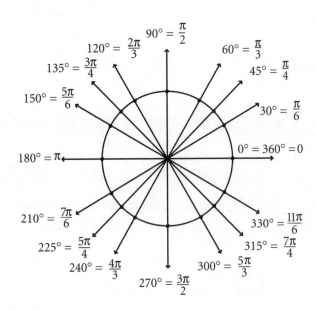

Figure 4.7

Solution

For this angle, we use a $\dfrac{\pi}{6}$, $\dfrac{\pi}{3}$, $\dfrac{\pi}{2}$ triangle, as shown in Figure 4.8 to find the *x*- and *y*-values. The *y*-value of the point where the ray of angle $\dfrac{2\pi}{3}$ hits the unit circle is $y = \dfrac{\sqrt{3}}{2}$. Thus, $\sin\left(\dfrac{2\pi}{3}\right) = \dfrac{\sqrt{3}}{2}$.

The x-value is negative, $x = -\dfrac{1}{2}$, so $\cos\left(\dfrac{2\pi}{3}\right) = -\dfrac{1}{2}$.

Figure 4.8

Example

Find the sine and cosine of $\dfrac{5\pi}{4} = 225°$.

Solution

Because $225°$ is a multiple of $45°$, we use a $45°, 45°, 90°$ triangle to find the x- and y-values. As seen in Figure 4.9, both coordinates are negative, so $\sin\left(\dfrac{5\pi}{4}\right) = -\dfrac{\sqrt{2}}{2}$ and $\cos\left(\dfrac{5\pi}{4}\right) = -\dfrac{\sqrt{2}}{2}$.

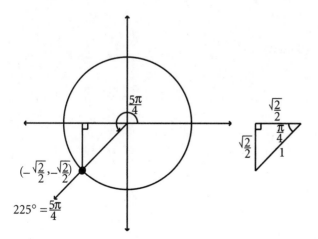

Figure 4.9

Example

Find all of the trigonometric values for $90° = \dfrac{\pi}{2}$.

Solution

Even though there isn't a triangle here, there is still a point on the unit circle. See Figure 4.10. We conclude that $\cos\left(\dfrac{\pi}{2}\right) = 0$ and $\sin\left(\dfrac{\pi}{2}\right) = 1$ from the x- and y-values of the point. Using the trigonometric identities, we can calculate that $\csc\left(\dfrac{\pi}{2}\right) = \dfrac{1}{\sin(\frac{\pi}{2})} = 1$ and $\cot\left(\dfrac{\pi}{2}\right) = \dfrac{\cos(\frac{\pi}{2})}{\sin(\frac{\pi}{2})} = \dfrac{0}{1} = 1$. The tangent and secant

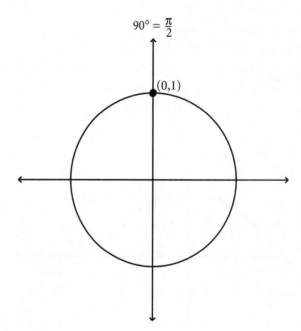

Figure 4.10

functions, however, involve division by 0 and thus are undefined. The angle $x = \dfrac{\pi}{2}$ is not in the domain of tan and sec.

Notice that all of the trigonometric functions are the same at $0° = 0$ and $360° = 2\pi$. This is because turning $360°$ leaves you facing in your original direction. Thus, everything repeats at this point.

Using the table along with the fact that everything repeats, we can sketch the graphs of $\sin(x)$ and $\cos(x)$. See Figures 4.11 and 4.12.

The functions sine and cosine are classic examples of periodic, or *oscillating*, functions.

Figure 4.11

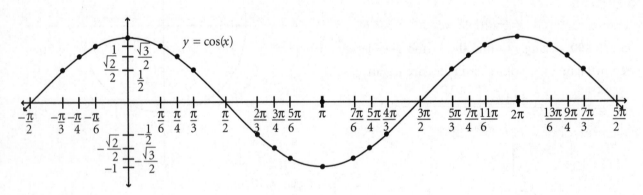

Figure 4.12

Practice

Use the unit circle and the trigonometric identities to complete the following table. Find the answers to questions 19 through 38.

	$\sin(x)$	$\cos(x)$	$\tan(x)$	$\sec(x)$	$\csc(x)$	$\cot(x)$
$0° = 0$	0	1	0	1	undef.	undef.
$30° = \dfrac{\pi}{6}$	$\dfrac{1}{2}$	$\dfrac{\sqrt{3}}{2}$	*19*	$\dfrac{2\sqrt{3}}{3}$	2	$\sqrt{3}$
$45° = \dfrac{\pi}{4}$	$\dfrac{\sqrt{2}}{2}$	$\dfrac{\sqrt{2}}{2}$	1	$\sqrt{2}$	$\sqrt{2}$	1
$60° = \dfrac{\pi}{3}$	$\dfrac{\sqrt{3}}{2}$	$\dfrac{1}{2}$	$\sqrt{3}$	2	$\dfrac{2\sqrt{3}}{3}$	$\dfrac{\sqrt{3}}{3}$
$90° = \dfrac{\pi}{2}$	1	0	undef.	undef.	1	0
$120° = \dfrac{2\pi}{3}$	$\dfrac{\sqrt{3}}{2}$	$-\dfrac{1}{2}$	$-\sqrt{3}$	-2	*20*	$-\dfrac{\sqrt{3}}{3}$
$135° = \dfrac{3\pi}{4}$	*21*	*22*	*23*	*24*	*25*	*26*
$150° = \dfrac{5\pi}{6}$	$\dfrac{1}{2}$	$-\dfrac{\sqrt{3}}{2}$	$-\dfrac{\sqrt{3}}{3}$	$-\dfrac{2\sqrt{3}}{3}$	2	$-\sqrt{3}$
$180° = \pi$	*27*	-1	0	-1	*28*	undef.
$210° = \dfrac{7\pi}{6}$	$-\dfrac{1}{2}$	$-\dfrac{\sqrt{3}}{2}$	$\dfrac{\sqrt{3}}{3}$	$-\dfrac{2\sqrt{3}}{3}$	-2	*29*
$225° = \dfrac{5\pi}{4}$	$-\dfrac{\sqrt{2}}{2}$	$-\dfrac{\sqrt{2}}{2}$	1	$-\sqrt{2}$	$-\sqrt{2}$	1
$240° = \dfrac{4\pi}{3}$	*30*	*31*	*32*	*33*	*34*	*35*
$270° = \dfrac{3\pi}{2}$	-1	0	undef.	undef.	-1	0
$300° = \dfrac{5\pi}{3}$	$-\dfrac{\sqrt{3}}{2}$	$\dfrac{1}{2}$	$-\sqrt{3}$	2	$-\dfrac{2\sqrt{3}}{3}$	$-\dfrac{\sqrt{3}}{3}$
$315° = \dfrac{7\pi}{4}$	$-\dfrac{\sqrt{2}}{2}$	$\dfrac{\sqrt{2}}{2}$	-1	*36*	$-\sqrt{2}$	-1
$330° = \dfrac{11\pi}{6}$	*37*	*38*	$-\dfrac{\sqrt{3}}{3}$	$\dfrac{2\sqrt{3}}{3}$	-2	$-\sqrt{3}$
$360° = 2\pi$	0	1	0	1	undef.	undef.

Note: The numbers appearing in bold with asterisks are questions 19 through 38.

19. Find the value that goes in the position in the table where you see *19*.

20. Find the value that goes in the position in the table where you see *20*.

21. Find the value that goes in the position in the table where you see *21*.

22. Find the value that goes in the position in the table where you see *22*.

23. Find the value that goes in the position in the table where you see *23*.

24. Find the value that goes in the position in the table where you see *24*.

25. Find the value that goes in the position in the table where you see *25*.

26. Find the value that goes in the position in the table where you see *26*.

27. Find the value that goes in the position in the table where you see *27*.

28. Find the value that goes in the position in the table where you see *28*.

29. Find the value that goes in the position in the table where you see *29*.

30. Find the value that goes in the position in the table where you see *30*.

31. Find the value that goes in the position in the table where you see *31*.

32. Find the value that goes in the position in the table where you see *32*.

33. Find the value that goes in the position in the table where you see *33*.

34. Find the value that goes in the position in the table where you see *34*.

35. Find the value that goes in the position in the table where you see *35*.

36. Find the value that goes in the position in the table where you see *36*.

37. Find the value that goes in the position in the table where you see *37*.

38. Find the value that goes in the position in the table where you see *38*.

Solving Simply Trigonometric Equations

The chart can be used to solve some simple equations.

Example
Find all values of x between 0 and 2π such that $\cos(x) = -\dfrac{\sqrt{2}}{2}$.

Solution
Note that multiples of $\dfrac{\pi}{4}$ have cosines equal to $\dfrac{\sqrt{2}}{2}$ or $-\dfrac{\sqrt{2}}{2}$. Of these, the values that solve the equation are $x = \dfrac{3\pi}{4}$ and $x = \dfrac{5\pi}{4}$.

Practice

For questions 39 and 40, find the value(s) of x between 0 and 2π that satisfy the given equation.

39. $\sin(x) = -\dfrac{\sqrt{3}}{2}$.

40. $\cos(x) = -1$.

5 ▶ LIMITS AND CONTINUITY

The notion of a *limit* is the single most important underlying concept upon which calculus is built. We can use the notion of a limit to describe the behavior of a function *near* a particular input, even when the function is not defined there.

Limits can be illustrated using graphs and tables of values. For example, consider the function whose graph is shown in Figure 5.1. We can't talk about $f(x)$ at $x = 2$ because of the unshaded circle on its graph. But, we *can* talk about what happens close to 2. The values of the function at x-values close to 2 are listed in the table.

x	$f(x)$
1.9	5.39
1.99	5.0399
1.999	5.003999
1.9999	5.000399999
↓	↓
2	???
↑	↑
2.0001	4.99959999
2.001	4.995999
2.01	4.9599
2.1	4.59

Figure 5.1

The domain of f is $x \neq 2$. We can't plug $x = 2$ into f. However, the hole appears to be at $(2,5)$. How do we know the hole has a y-value of 5? Well, the points on the curve with x-values *near* $x = 2$ have y-values *close* to $y = 5$. The closer we get to $x = 2$, the closer the y-values of the points come to $y = 5$.

The mathematical shorthand for this is $\lim\limits_{x \to 2} f(x) = 5$, which is read as "the limit as x approaches 2 of $f(x)$ is 5."

The utility of limits lies in the fact that f need not be defined *at* the value a in order to have a limit as x approaches a.

We can also approach points from either the left or from the right. For example, consider the graph of $y = g(x)$ in Figure 5.2.

Here, $\lim\limits_{x \to 1^-} g(x) = 4$ and $\lim\limits_{x \to 1^+} g(x) = 2$. The little minus in $\lim\limits_{x \to 1^-}$ means that we approach $x = 1$ using numbers less than (to the left) of $x = 1$. As we approach $x = 1$ from the left-hand side, we slide up the graph through y-values that approach 4. Similarly, the plus in $\lim\limits_{x \to 1^+}$ means "approach from the right."

From the right, the height of the graph slides down to $y = 2$ as x approaches 1.

In this example, $\lim\limits_{x \to 1} g(x)$ does not exist because there is no single y-value to which all of the points near $x = 1$ get close. Some are close to 4, and others are close to 2. Because there is no agreement, there is no limit.

As another example, consider the graph of $y = h(x)$ in Figure 5.3. Here, $\lim\limits_{x \to 3^-} h(x) = 2$ because sliding up to $x = 3$ from the left has us pass through points with y-values near 2. Similarly, $\lim\limits_{x \to 3^+} h(x) = 2$. Because there *is* agreement from the left and right, we have the general limit, $\lim\limits_{x \to 3} h(x) = 2$. Notice that what happens exactly at $x = 3$ is irrelevant. Here $h(3) = 5$, but the resulting point at $(3,5)$ has no bearing on the limit as x approaches 3.

Vertical asymptotes correspond with infinite limits. For example, consider the graph of $y = k(x)$ in Figure 5.4.

Figure 5.2

Figure 5.3

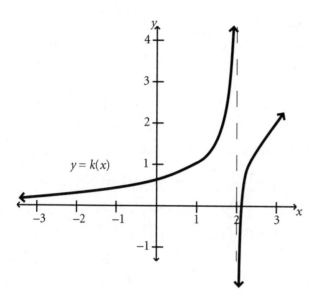

Figure 5.4

The difference from the limits discussed previously is that now, in each case, the function is *continuous* at the point where we are computing the limit, meaning that there are no holes, jumps, or asymptotes there. Symbolically, we say f is continuous at $x = a$ if

$$\lim_{x \to a^-} f(x) = \lim_{x \to a^+} f(x) = f(a).$$

Practice

Use the graphs in Figure 5.5 to evaluate the following.

1. $\lim_{x \to -1^-} f(x)$

2. $\lim_{x \to -1^+} f(x)$

3. $\lim_{x \to -1} f(x)$

4. $f(-1)$

5. Is f continuous at $x = -1$?

6. $\lim_{x \to 3^-} f(x)$

Here, we write $\lim_{x \to 2^-} k(x) = \infty$ and $\lim_{x \to 2^+} k(x) = -\infty$. These statements simply suggest what the graph is doing on either side of 2. The limits technically do not exist since they are infinite.

Look back at the graphs in this lesson. Would you agree with the following?

$$\lim_{x \to 0} f(x) = 9, \quad \lim_{x \to -3} g(x) = 0, \quad \lim_{x \to 4} h(x) = 1$$

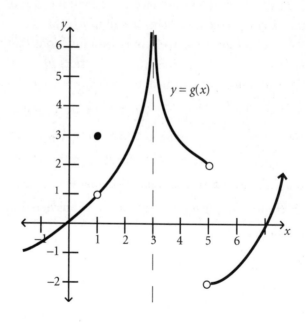

Figure 5.5

7. $\lim\limits_{x \to 3^+} f(x)$

8. $\lim\limits_{x \to 3} f(x)$

9. $f(3)$

10. Is f continuous at $x = 1$?

11. $\lim\limits_{x \to 1} g(x)$

12. $g(1)$

13. $\lim\limits_{x \to 3^-} g(x)$

14. $\lim\limits_{x \to 3} g(x)$

15. $\lim\limits_{x \to 5^-} g(x)$

16. $\lim\limits_{x \to 5^+} g(x)$

Evaluating Limits Algebraically

It is not necessary to have the graph of a function to evaluate its limits. For instance, if f is continuous at a, then its limit as x approaches a is simply $f(a)$.

Technically, this works only with functions that are *polynomials* like $4x^5 - 10x^3 - 7$, roots like $\sqrt{x}$, *rational functions* (formed by dividing two polynomials) like $\dfrac{3x - 5}{2x^3 + x^2 + 1}$, trigonometric functions, and *transcendental functions* like $\ln(x)$, and e^x. Because this works for any combination of these functions added, subtracted, multiplied, divided, or composed, it works also for every function considered in this book.

Example

Evaluate $\lim\limits_{x \to 4} \dfrac{x + 5}{x^2 + 10x}$ and $\lim\limits_{x \to -2} (3x^2 + x - 7)$.

Solution

Because 4 can be plugged into $\dfrac{x + 5}{x^2 + 10x}$ without there being a division by zero, the limit $\lim\limits_{x \to 4} \dfrac{x + 5}{x^2 + 10x}$ $= \dfrac{4 + 5}{16 + 40} = \dfrac{9}{56}$. Similarly, $\lim\limits_{x \to -2} (3x^2 + x - 7) = 3(-2)^2 + (-2) - 7 = 3$.

Practice

Evaluate the following limits.

17. $\lim\limits_{x \to 1} (10x^3 + 4x^2 - 5x + 7)$

18. $\lim\limits_{x \to 3} \dfrac{x - 3}{x^2 + x}$

19. $\lim\limits_{x \to 2} \left(\dfrac{x^3 - 4}{10x + 3} + x^2 \right)$

20. $\lim\limits_{x \to \frac{\pi}{6}} \dfrac{\sin(x)}{x}$

21. $\lim\limits_{a \to 0} (2x + a + 1)$

22. $\lim\limits_{x \to -2} e^{-3x^2}$

Dividing by a tiny number is equivalent to multiplying by an enormous number. For example:

$$5 \div \dfrac{1}{10,000} = 5 \cdot \dfrac{10,000}{1} = 50,000$$

It is for this reason that if the denominator of a fraction approaches zero while the numerator goes to a nonzero number, the result is an infinite limit.

A classic example is $f(x) = \dfrac{1}{x}$ (graphed in Figure 5.6).

$y = \dfrac{1}{x}$

Figure 5.6

Because the denominator goes to zero while the numerator stays one in all of these cases, there is a vertical asymptote at $x = 0$. The function therefore approaches either positive or negative infinity from either side. When x is less than zero, as it always is when $x \to 0^-$, the function $\dfrac{1}{x}$ is also negative. Thus, $\lim\limits_{x \to 0^-} \dfrac{1}{x} = -\infty$. Similarly, as $x \to 0^+$, $\dfrac{1}{x}$ is always positive, so $\lim\limits_{x \to 0^+} \dfrac{1}{x} = \infty$. Finally, because the limit from the two sides are different, the undirected limit $\lim\limits_{x \to 0} \dfrac{1}{x}$ does not exist.

Example

Evaluate $\lim\limits_{x \to 2^+} \dfrac{x+3}{(x-2)(x-4)}$.

Solution

The numerator approaches 5 while the denominator approaches 0 as x approaches 2 from the right. Therefore, this limit from the right is either ∞ or $-\infty$. What we need to determine is whether the function is positive or negative at x-values just slightly larger than 2. We do this by looking at each factor individually.

As $x \to 2^+$, the values we are plugging into each factor are slightly larger than 2. So, $(x+3)$ and $(x-2)$ are both positive, while $(x-4)$ is negative. Because the function $\dfrac{x+3}{(x-2)(x-4)}$ is made of two positive parts and one negative part, the entire fraction will be negative. Thus, $\lim\limits_{x \to 2^+} \dfrac{x+3}{(x-2)(x-4)} = -\infty$.

Because this is negative, the limit is $-\infty$. Another method will be covered in Lesson 13.

Example

Evaluate $\lim\limits_{x \to -3^-} \dfrac{(x+1)(2-x)}{(x+3)(x+5)}$.

Solution

Here, the numerator approaches -10, which isn't zero, while the denominator approaches zero, so the limit is either ∞ or $-\infty$. As $x \to -3^-$, the values we are plugging into each factor are slightly smaller than -3. So, $(x+1)$ and $(x+3)$ are both negative, while $(2-x)$ and $(x+5)$ are both positive.

The combination of two negative factors and two positive factors is positive, thus:

$$\lim\limits_{x \to -3^-} \dfrac{(x+1)(2-x)}{(x+3)(x+5)} = \infty$$

Practice

Evaluate the following limits.

23. $\lim\limits_{x \to 1^-} \dfrac{1}{x - 1}$

24. $\lim\limits_{x \to 4^+} \dfrac{x + 5}{x - 4}$

25. $\lim\limits_{x \to 3^-}$

26. $\lim\limits_{x \to 3^+} \dfrac{(x + 2)(x - 5)}{(x + 6)(x - 3)}$

27. $\lim\limits_{x \to 2} \dfrac{(x + 5)(x - 5)}{(x - 3)(x + 4)}$

28. $\lim\limits_{x \to -5} \dfrac{x - 2}{(x + 5)^2}$

When both the numerator *and* the denominator go to zero, there are some common tricks for simplifying the limit. The first is to factor and cancel. The second is to rationalize. These are illustrated next.

Example

Evaluate $\lim\limits_{x \to 4} \dfrac{x^2 - 2x - 8}{x^2 + x - 20}$.

Solution

Here, *both* the numerator and denominator go to zero, so we aren't guaranteed an infinite limit. First, factor the numerator and denominator.

$$\lim\limits_{x \to 4} \dfrac{x^2 - 2x - 8}{x^2 + x - 20} = \lim\limits_{x \to 4} \dfrac{(x - 4)(x + 2)}{(x - 4)(x + 5)}$$

Because $x \neq 4$ as $x \to 4$, we can cancel $\dfrac{x - 4}{x - 4} = 1$.

$$\lim\limits_{x \to 4} \dfrac{x^2 - 2x - 8}{x^2 + x - 20} =$$

$$\lim\limits_{x \to 4} \dfrac{(x - 4)(x + 2)}{(x - 4)(x + 5)} = \lim\limits_{x \to 4} \dfrac{(x + 2)}{(x + 5)}$$

Now we can plug in without dividing by zero.

$$\lim\limits_{x \to 4} \dfrac{x^2 - 2x - 8}{x^2 + x - 20} = \lim\limits_{x \to 4} \dfrac{(x + 2)}{(x + 5)} = \dfrac{6}{9} = \dfrac{2}{3}$$

The following example utilizes the trick of rationalizing.

Example

Evaluate $\lim\limits_{x \to 9} \dfrac{\sqrt{x} - 3}{x - 9}$.

Solution

Because both numerator and denominator go to zero, a trick is necessary. First, multiply the top and bottom by the part with the square root, but with the opposite sign between them.

$$\lim\limits_{x \to 9} \dfrac{\sqrt{x} - 3}{x - 9} = \lim\limits_{x \to 9} \left(\dfrac{\sqrt{x} - 3}{x - 9} \right) \cdot \left(\dfrac{\sqrt{x} + 3}{\sqrt{x} + 3} \right)$$

Simplify.

$$\lim\limits_{x \to 9} \dfrac{\sqrt{x} - 3}{x - 9} = \lim\limits_{x \to 9} \dfrac{x + 3\sqrt{x} - 3\sqrt{x} - 9}{(x - 9)(\sqrt{x} + 3)}$$

$$= \lim\limits_{x \to 9} \dfrac{x - 9}{(x - 9)(\sqrt{x} + 3)}$$

Eliminate $\dfrac{x - 9}{x - 9} = 1$.

$$\lim_{x \to 9} \frac{\sqrt{x} - 3}{x - 9} = \lim_{x \to 9} \frac{(x - 9)}{(x - 9)(\sqrt{x} + 3)}$$

Plug in.

$$\lim_{x \to 9} \frac{\sqrt{x} - 3}{x - 9} = \lim_{x \to 9} \frac{1}{(\sqrt{x} + 3)} = \frac{1}{6}$$

Factoring tricks can even be useful when dealing with transcendental functions.

Example
Evaluate $\displaystyle\lim_{x \to 0^+} \dfrac{1 - e^x}{1 - e^{2x}}$

Solution
To compute this limit, use the fact that $e^{2x} = \left(e^x\right)^2$ and factor the denominator as a difference of squares. Then, cancel factors that are common to both the numerator and denominator, and substitute $x = 0$ into the simplified expression, as follows:

$$\lim_{x \to 0^+} \frac{1 - e^x}{1 - e^{2x}} = \lim_{x \to 0^+} \frac{1 - e^x}{1 - \left(e^x\right)^2} = \lim_{x \to 0^+} \frac{1 - e^x}{\left(1 - e^x\right)\left(1 + e^x\right)}$$

$$\lim_{x \to 0^+} \frac{1}{1 + e^x} = \frac{1}{1 + e^0} = \frac{1}{1 + 1} = \frac{1}{2}$$

Practice

Evaluate the following limits.

29. $\displaystyle\lim_{x \to -2^+} \dfrac{(x - 6)(x + 2)}{(x + 2)(x + 1)}$

30. $\displaystyle\lim_{x \to 2} \dfrac{x - 2}{x^2 - 4}$

31. $\displaystyle\lim_{x \to 4} \dfrac{x^2 - 9}{x + 3}$

32. $\displaystyle\lim_{x \to 3} \dfrac{x^2 - 4x + 3}{x^2 + 2x - 15}$

33. $\displaystyle\lim_{x \to 3^-} \dfrac{x + 5}{x - 3}$

34. $\displaystyle\lim_{x \to 25} \dfrac{\sqrt{x} - 5}{(x - 25)(x + 1)}$

35. $\displaystyle\lim_{a \to 0} \dfrac{(x + a)^2 - x^2}{a}$

36. $\displaystyle\lim_{a \to 0} \dfrac{a}{\sqrt{x + a} - \sqrt{x}}$

37. $\displaystyle\lim_{x \to \pi} \dfrac{\cos(x) + 1}{\cos(x) - 1}$

38. $\displaystyle\lim_{h \to 0} \dfrac{2(x + h)^3 - 2x^3}{h}$

39. $\displaystyle\lim_{x \to 3} \dfrac{x^3 + 2x^2 - 15x}{x^2 - 9}$

40. $\displaystyle\lim_{z \to \ln(2)} \dfrac{\left(e^z\right)^2 - 6\left(e^z\right) + 8}{2 - e^z}$

6 ▶ DERIVATIVES

LESSON

Straight lines are convenient to deal with, but most functions have curved graphs. This does not stop us from projecting straight lines on them! For example, at the point marked *x* on the graph in Figure 6.1, the function is clearly increasing. However, exactly how fast is the function increasing at that point? Since "how fast" refers to a slope, we draw in the *tangent line,* the line straight through the point that heads in the same direction as the curve (see Figure 6.2). The slope of the tangent line tells us how fast the function is increasing at the given point.

Figure 6.1

Figure 6.2

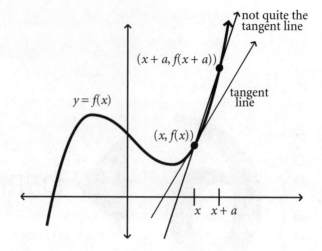

Figure 6.3

We can figure out the y-value of this point by plugging x into f and getting $(x, f(x))$. However, we can't get the slope of the tangent line when we have just one point. To get a second point, we go ahead a little further along the graph (see Figure 6.3). If we go ahead by distance a, the second point will have an x-value of $x + a$ and a y-value of $f(x + a)$.

Because this second point is on the curve and not on the tangent line, we get a line that is not quite the tangent line. Still, its slope will be close to the one we want, so we calculate as follows:

$$slope = \frac{f(x + a) - f(x)}{(x + a) - x} = \frac{f(x + a) - f(x)}{a}$$

To make things more accurate, we pick a second point that is closer to the original point $(x, f(x))$ by using a smaller a. This is depicted in Figure 6.4.

In fact, if we take the limit as a goes to zero from the right, we will get the slope of the tangent line exactly. The situation is completely similar if we take $a < 0$ so that $x + a$ is to the *left* of x. This process gives us the *derivative* of $f(x)$ and is written $f'(x)$.

$$f'(x) = \lim_{a \to 0} \frac{f(x + a) - f(x)}{a}$$

$$= \text{slope of the tangent line at point } (x, f(x))$$

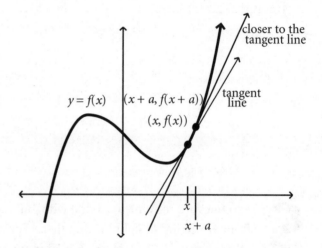

Figure 6.4

Example
What is the derivative of $f(x) = x^2$?

Solution
Start with the definition of the derivative.

$$f'(x) = \lim_{a \to 0} \frac{f(x + a) - f(x)}{a}$$

⚠ CAUTION
Remember, $f(x + a) \neq f(x) + f(a)$.

Use $f(x) = x^2$.

$$f'(x) = \lim_{a \to 0} \frac{(x + a)^2 - x^2}{a}$$

Multiply out and simplify.

$$f'(x) = \lim_{a \to 0} \frac{\cancel{x^2} + 2ax + a^2 - \cancel{x^2}}{a}$$

Factor and simplify.

$$f'(x) = \lim_{a \to 0} \frac{(2x + a)\cancel{a}}{\cancel{a}}$$

Evaluate the limit.

$$f'(x) = \lim_{a \to 0} 2x + a = 2x$$

The derivative is $f'(x) = 2x$. This means that the slope at any point on the curve $y = x^2$ is exactly twice the x-coordinate. The situation at $x = -2$, $x = 0$, and $x = 1$ is shown in Figure 6.5.

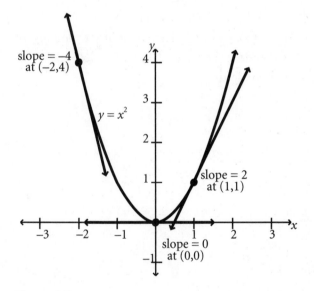

slope = −4 at (−2,4)

$y = x^2$

slope = 2 at (1,1)

slope = 0 at (0,0)

Figure 6.5

Example
What is the slope of the line tangent to $g(x) = \sqrt{x}$ at $x = 9$?

Solution
Start with the definition of the derivative.

$$g'(x) = \lim_{a \to 0} \frac{g(x + a) - g(x)}{a}$$

Use $g(x) = \sqrt{x}$.

$$g'(x) = \lim_{a \to 0} \frac{\sqrt{x + a} - \sqrt{x}}{a}$$

Rationalize the numerator.

$$g'(x) = \lim_{a \to 0} \left(\frac{\sqrt{x + a} - \sqrt{x}}{a} \right)\left(\frac{\sqrt{x + a} + \sqrt{x}}{\sqrt{x + a} + \sqrt{x}} \right)$$

Multiply and simplify.

$$g'(x) = \lim_{a \to 0} \frac{\cancel{x} + a + \cancel{\sqrt{x} \cdot \sqrt{x + a}} - \cancel{\sqrt{x} \cdot \sqrt{x + a}} - \cancel{x}}{a(\sqrt{x + a} + \sqrt{x})}$$

Simplify.

$$g'(x) = \lim_{a \to 0} \frac{\cancel{a}}{\cancel{a}(\sqrt{x + a} + \sqrt{x})}$$

Plug in to evaluate the limit.

$$g'(x) = \lim_{a \to 0} \frac{1}{\sqrt{x + a} + \sqrt{x}}$$

$$= \frac{1}{\sqrt{x + 0} + \sqrt{x}}$$

$$= \frac{1}{2\sqrt{x}}$$

The derivative of $g(x) = \sqrt{x}$ is thus $g'(x) = \frac{1}{2\sqrt{x}}$. This means that at $x = 9$, the slope of the tangent line is $g'(9) = \frac{1}{2\sqrt{9}} = \frac{1}{6}$. This is illustrated in Figure 6.6.

Example
Find the equation of the tangent line to $h(x) = 2x^2 - 5x + 1$ at $x = 3$.

Figure 6.6

Solution

To find the equation of the tangent line, we need a point and a slope. The y-value at $x = 3$ is $h(3) = 2(3)^2 - 5(3) + 1 = 4$, so the point is $(3,4)$. And to get the slope, we need the derivative. Start with the definition of the derivative.

$$h'(x) = \lim_{a \to 0} \frac{h(x + a) - h(x)}{a}$$

Use $h(x) = 2x^2 - 5x + 1$.

$$h'(x) = \lim_{a \to 0} \frac{2(x + a)^2 - 5(x + a) + 1 - (2x^2 - 5x + 1)}{a}$$

Multiply out and simplify.

$$h'(x) = \lim_{a \to 0} \frac{2x^2 + 4ax + 2a^2 - 5x - 5a + 1 - 2x^2 + 5x - 1}{a}$$

Factor out and simplify.

$$h'(x) = \lim_{a \to 0} \frac{(4x + 2a - 5)a}{a}$$

Evaluate the limit.

$$h'(x) = \lim_{a \to 0}(4x + 2a - 5) = 4x - 5$$

Thus, the derivative of $h(x) = 2x^2 - 5x + 1$ is $h'(x) = 4x - 5$. The slope at $x = 3$ is $h'(3) = 4(3) - 5 = 7$. The equation of the tangent line is therefore $y = 7(x - 3) + 4 = 7x - 17$. This is shown in Figure 6.7.

The *absolute value of* x, denoted $|x|$, tells you how far x is from zero. For instance, $|5| = 5$ (since 5 is five units from 0 on the right-hand side) and $|-4| = 4$ (since –4 is four units from 0, just on the left-hand side). Symbolically,

$$|x| = \begin{cases} x, \text{ whenever } x \geq 0 \\ -x, \text{ whenever } x < 0 \end{cases}$$

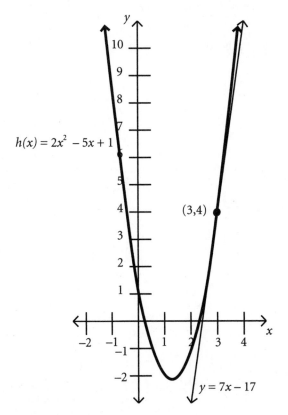

$h(x) = 2x^2 - 5x + 1$

$(3,4)$

$y = 7x - 17$

Figure 6.7

Practice

1. Find the derivative of $f(x) = 8x + 2$.

2. If $h(x) = x^2 + 5$, then what is $h'(x)$?

3. Find the derivative of $g(x) = 10$.

4. Find the derivative of $f(x) = 3\sqrt{x}$.

5. If $k(x) = x^3$, then what is $k'(x)$?

6. Find the slope of $f(x) = 3x^2 + x$ at $x = 2$.

7. Find the slope of $g(x) = 3\sqrt{x}$ at $x = 16$.

8. Where does the graph of $g(x) = x^2 - 4x + 1$ have a slope of 0?

9. Find the equation of the tangent line to $h(x) = 1 - x^2$ at $(2, -3)$.

10. Find the equation of the tangent line to $k(x) = 5x^2 + 2x$ when $x = 1$.

Derivatives Don't Always Exist!

Because the definition of $f'(x)$ involves a full-blown limit in order for it to exist, the following left- and right-hand limits must be equal:

$$\lim_{a \to 0^-} \frac{f(x+a) - f(x)}{a} = \lim_{a \to 0^+} \frac{f(x+a) - f(x)}{a}$$

This is not always the case, though.

Example

Let us try to compute the derivative of $A(x) = |x|$ at $x = 0$ (shown in Figure 6.8).

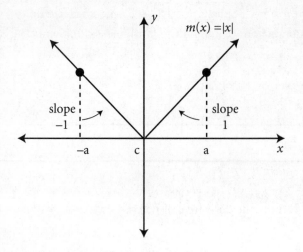

For a > 0, we have

$$\frac{m(0+a)-m(0)}{a} = \frac{|a|-|0|}{a} = \frac{a}{a} = 1$$

So, $\lim_{a\to 0^+} \frac{m(0+a)-m(0)}{a} = \lim_{a\to 0^+} 1 = 1.$

Now, let us take $a < 0$. Doing so, we get:

$$\frac{m(0+a)-m(0)}{a} = \frac{|a|-|0|}{a} = \frac{-a}{a} = -1$$

So, $\lim_{a\to 0^-} \frac{m(0+a)-m(0)}{a} = \lim_{a\to 0^-} (-1) = -1.$

But, the left- and right-hand limits aren't equal. So, the full-blown limit $\lim_{a\to 0}\frac{m(0+a)-m(0)}{a}$ does not exist. Thus, $m'(0)$ doesn't exist. Geometrically, the *sharp corner* in the graph of $y = m(x)$ at $x = 0$ is the reason the derivative doesn't exist there.

7 ▶ BASIC RULES OF DIFFERENTIATION

Using the limit definition to find derivatives can be very tedious. Luckily, there are many shortcuts available. For example, if function f is a constant, like $f(x) = 5$ or $f(x) = 18$, then $f'(x) = 0$. This can be proven for all constants c at the same time in the following manner.

If:

$$f(x) = c$$

then:

$$f'(x) = \lim_{a \to 0} \frac{f(x + a) - f(x)}{a} = \lim_{a \to 0} \frac{c - c}{a} = \lim_{a \to 0} \frac{0}{a} = 0$$

All of the general rules in this chapter can be proven in such a manner, using the limit definition of the derivative, though we shall not actually do so. The first rule is the Constant Rule, which says that if $f(x) = c$ where c is a constant, then $f'(x) = 0$.

Before we go any further, a word needs to be said about notation. The concept of the derivative was discovered by both Isaac Newton and Gottfried Leibniz. Newton would put a dot over a quantity to represent its derivative, much like we have used the prime notation $f'(x)$ to represent the derivative of $f(x)$. Leibniz would write the derivative of y (where x is the variable) as $\frac{dy}{dx}$. Newton's notation is certainly more

CONSTANT RULE

If $f(x) = c$ where c is a constant, then $f'(x) = 0$.

And, using Leibniz's notation, if c is a constant, then $\dfrac{d}{dx}(c) = 0$.

POWER RULE

$$\frac{d}{dx}(x^n) = n \cdot x^{n-1}$$

convenient, but Leibniz's enables us to represent "take the derivative of something" as $\dfrac{d}{dx}$ (something).

Thus, if $y = f(x)$, then $\dfrac{dy}{dx} = \dfrac{d}{dx}(f(x)) = f'(x)$. Using Leibniz's notation, the Constant Rule where c is a constant is expressed as $\dfrac{d}{dx}(c) = 0$.

The next rule is the Power Rule, which is stated: $\dfrac{d}{dx}(x^n) = n \cdot x^{n-1}$. This rule says "multiply the exponent in front and then subtract one from it."

Example

Differentiate $f(x) = x^2$.

Solution

$f'(x) = 2x^{2-1} = 2x^1 = 2x$

Example

Differentiate $y = x^8$.

Solution

$\dfrac{dy}{dx} = 8x^{8-1} = 8x^7$

Example

Differentiate $g(x) = \sqrt{x}$.

Solution

To use the Power Rule, we need $g(x)$ expressed as x raised to a power, or:

$$g(x) = x^{\frac{1}{2}}$$

$$g'(x) = \frac{1}{2}x^{\frac{1}{2}-1} = \frac{1}{2}x^{-\frac{1}{2}} = \frac{1}{2} \cdot \frac{1}{\sqrt{x}} = \frac{1}{2\sqrt{x}}$$

Notice how much easier it is to use the Power Rule to compute the derivative than it was using the limit definition of the derivative in Lesson 6.

THE CONSTANT COEFFICIENT RULE

If a function has a constant multiplied in front, leave it while you take the derivative of the rest. Using Leibniz's notation, $\frac{d}{dx}(cf(x)) = c\frac{d}{dx}(f(x))$, where c is a constant.

Example

Differentiate $y = \frac{1}{x^2}$.

Solution

First, rewrite y as x^{-2} so that it becomes x raised to a power. Then,

$$\frac{dy}{dx} = \frac{d}{dx}\left(\frac{1}{x^2}\right)$$

$$= \frac{d}{dx}(x^{-2}) = -2x^{-2-1} = -2x^{-3} = \frac{-2}{x^3}$$

Notice that $\frac{d}{dt}$ means "take the derivative with respect to variable t." While x is often used as the variable, so the derivative of $y = f(x)$ is $\frac{dy}{dx} = f'(x)$, sometimes it is convenient to use other variables. If $y = f(u)$, then $\frac{dy}{du} = f'(u)$ is the derivative of f with respect to u, for example.

Example

Differentiate $y = \sqrt[3]{t}$.

Solution

$$\frac{d}{dt}(\sqrt[3]{t}) = \frac{d}{dt}(t^{\frac{1}{3}}) = \frac{1}{3}t^{\frac{1}{3}-1} = \frac{1}{3}t^{-\frac{2}{3}} = \frac{1}{3t^{\frac{2}{3}}}$$

Example

Differentiate $y = \frac{\sqrt{t}}{t^3}$.

Solution

Rewrite using the exponent rules.

$$y = \frac{\sqrt{t}}{t^3} = \frac{t^{\frac{1}{2}}}{t^3} = t^{\frac{1}{2}} \cdot t^{-3} = t^{-\frac{5}{2}}$$

Now, use the Power Rule to differentiate.

$$\frac{dy}{dt} = -\frac{5}{2}t^{-\frac{7}{2}}$$

Practice

Differentiate each of the following.

1. $f(x) = x^5$

2. $y = x^{21}$

3. $g(u) = u^{-5}$

4. $h(x) = 8$

5. $y = t^{12}$

6. $y = x^{\frac{7}{5}}$

7. $f(x) = x^{100}$

8. $f(t) = -11$

9. $g(x) = x^{-\frac{4}{5}}$

10. $k(x) = \sqrt[4]{x}$

11. $y = \sqrt{u}$

12. $y = \frac{1}{x}$

13. $f(x) = \frac{1}{\sqrt{x}}$

14. $g(x) = \dfrac{1}{x\sqrt{x}}$

15. $h(t) = \dfrac{t^5}{t^{-2}}$

16. $y = \dfrac{\sqrt[3]{t}}{\sqrt[4]{t}}$

The Constant Coefficient Rule

The Constant Coefficient Rule is stated as follows: If a function has a constant multiplied in front, leave it while you take the derivative of the rest. This means that because $\dfrac{d}{dx}(x^8) = 8x^7$, the derivative of $5x^8$ would be $5 \cdot (8x^7) = 40x^7$. Just imagine that the constant steps aside and waits while you differentiate the rest.

Examples

Differentiate the following.

$f(x) = 11x^4$

$y = 10x^2$

$g(x) = 3\sqrt{x} = 3x^{\frac{1}{2}}$

$h(t) = \dfrac{4}{t^6} = 4t^{-6}$

$y = 12x$

$k(u) = \dfrac{15\sqrt[3]{u}}{4} = \dfrac{15}{4}u^{\frac{1}{3}}$

$A(r) = \pi r^2$

Solutions

$f'(x) = 11 \cdot (4x^3) = 44x^3$

$\dfrac{dy}{dx} = 10 \cdot (2x) = 20x$

$g'(x) = 3 \cdot \left(\dfrac{1}{2}x^{-\frac{1}{2}}\right) = \dfrac{3}{2}x^{-\frac{1}{2}} = \dfrac{3}{2\sqrt{x}}$

$h'(t) = 4 \cdot \left(-6t^{-7}\right) = -24t^{-7} = -\dfrac{24}{t^7}$

$\dfrac{dy}{dx} = 12(1) = 12$

$k'(u) = \dfrac{15}{4} \cdot \left(\dfrac{1}{3}u^{-\frac{2}{3}}\right) = \dfrac{5}{4}u^{-\frac{2}{3}} = \dfrac{5}{4u^{\frac{2}{3}}}$

$A'(r) = \pi \cdot (2r) = 2\pi r$

In that last example problem, don't forget that π is a constant, and thus $2\pi r$ should be treated just as you would $20r$ or $712r$.

> **⚠ CAUTION**
>
> Don't make the mistake that
> $$\dfrac{d}{dx}(c \cdot f(x)) = \underbrace{\dfrac{d}{dx}(c)}_{= 0} \cdot \dfrac{d}{dx}(f(x)) = 0.$$

The Additive Rule

Next, we will examine the Additive Rule, which says that if parts of a function are added together, differentiate the parts separately and add the results. We know that $\dfrac{d}{dx}(10x^2) = 20x$ and $\dfrac{d}{dx}(12x) = 12$. The Additive Rule then says that if $y = 10x^2 + 12x$, then $\dfrac{dy}{dx} = \dfrac{d}{dx}(10x^2 + 12x) = \dfrac{d}{dx}(10x^2) + \dfrac{d}{dx}(12x) = 20x + 12$.

Example

Differentiate $f(x) = 4x^5 + 30x^2$.

Solution

$f'(x) = \dfrac{d}{dx}(4x^5) + \dfrac{d}{dx}(30x^2) = 20x^4 + 60x$

If parts of a function are added together, differentiate the parts separately. Then, add the results. Using Leibniz's notation, $\frac{d}{dx}(f(x) + g(x)) = \frac{d}{dx}(f(x)) + \frac{d}{dx}(g(x))$.

Example
Differentiate $g(x) = x^3 - 4x^2$.

Solution
This can be rewritten as a sum:

$$g(x) = x^3 + (-4)x^2$$

thus:

$$g'(x) = \frac{d}{dx}\left(x^3\right) + \frac{d}{dx}\left(-4x^2\right) = 3x^2 + (-4) \cdot 2x$$
$$= 3x^2 - 8x.$$

The previous example shows that the Additive Rule applies to cases of subtraction as well.

Examples
Differentiate the following.

$$y = \sqrt{x} + 4 = x^{\frac{1}{2}} + 4$$

$$h(x) = 8x^5 + 10x^4 - 3x^3 + 7x^2 - 5x + 4$$

$$k(t) = 3t^{\frac{4}{5}} + \frac{2}{t} + e = 3t^{\frac{4}{5}} + 2t^{-1} + e$$

⚠ CAUTION
Remember, e is a number (approximately equal to 2.71828…).

Solutions

$$\frac{dy}{dx} = \frac{1}{2\sqrt{x}} + 0 = \frac{1}{2\sqrt{x}}$$

$$h'(x) = 40x^4 + 40x^3 - 9x^2 + 14x - 5$$

$$k'(t) = \frac{12}{5}t^{-\frac{1}{5}} - 2t^{-2} + 0 = \frac{12}{5\sqrt[5]{t}} - \frac{2}{t^2}$$

Practice

Differentiate the following.

17. $y = 6x^7$

18. $f(x) = \dfrac{-3}{x^{10}}$

19. $V(r) = \frac{4}{3}\pi r^3$

20. $g(t) = \dfrac{12t^4}{5}$

21. $k(x) = 1 - x^2$

22. $4t^3 - 3t^2 + 70$

23. $f(x) = 8x^3 + 3x^2$

24. $y = x^{-4} - 2x^{-3} - x$

25. $s(t) = \pi t^3 + et^2 + \sqrt{3}\,t + \ln(3)$

26. $F(x) = 6x^{100} + 10x^{50} - 4x^{25} + 2x^{10} - 9$

27. $g(x) = 3x^{\frac{1}{6}} + 5x^3$

28. $h(u) = u^5 + 4u^4 - 7u^3 - 2u^2 + 8u - 2$

29. $y = 3 + \dfrac{2}{x} + \dfrac{1}{x^2}$

30. $y = u^2 - u^{-2}$

31. $y = 4\sqrt{x} + 9 = 9\sqrt[3]{x}$

32. $f(x) = \dfrac{3}{2x^{-4}} - \dfrac{5}{x\sqrt{x}}$

The derivative of the derivative is called the *second derivative*. The derivative of that is the *third derivative*, and so on. Using notation, if $y = f(x)$, then the derivative is $\dfrac{dy}{dx} = f'(x)$, the second derivative is $\dfrac{d^2y}{dx^2} = f''(x)$, the third derivative is $\dfrac{d^3y}{dx^3} = f'''(x)$, and the tenth derivative, for example, is $\dfrac{d^{10}y}{dx^{10}} = f^{(10)}(x)$. We put the 10 in parentheses because counting the ten primes in $f''''''''''(x)$ is ridiculous.

Example

Find the first three derivatives of $y = \sqrt{x}$.

Solution

$$y = x^{\frac{1}{2}}$$

$$\frac{dy}{dx} = \frac{1}{2}x^{-\frac{1}{2}}$$

$$\frac{d^2y}{dx^2} = -\frac{1}{4}x^{-\frac{3}{2}}$$

$$\frac{d^3y}{dx^3} = \frac{3}{8}x^{-\frac{5}{2}}$$

When working on multiple derivatives like this, it makes sense to leave the exponents negative and fractional until all derivatives have been computed.

Example

Find all the derivatives of $f(x) = x^3 - 4x^2 + 5x - 7$.

Solution

$$f(x) = x^3 - 4x^2 + 5x - 7$$

$$f'(x) = 3x^2 - 8x + 5$$

$$f''(x) = 6x - 8$$

$$f'''(x) = 6$$

$$f''''(x) = 0$$

All of the subsequent derivatives will also be zero, so we can write

$$f^{(n)}(x) = 0 \text{ for } n \geq 4.$$

Practice

33. Find the first four derivatives of $f(x) = \dfrac{1}{x}$.

34. Find the second derivative of $s(t) = -16t^2 + at + b$, where a and b are constants.

35. Find the third derivative of $y = 3x^{-3} - 2x^{-2} + x^{-1}$.

36. Find the first three derivatives of $y = 6\sqrt[3]{t}$.

LESSON

8 ▶ RATES OF CHANGE

t is useful to contemplate slopes in practical situations. For example, suppose the following graph in Figure 8.1 is for $y = f(x)$, a function that gives the price y for various amounts x of cheese. Because the straight line goes through the points (1 lb.,\$2) and (2 lbs.,\$4), the slope $= \dfrac{\$4 - \$2}{2 \text{ lbs.} - 1 \text{ lb.}} = \dfrac{\$2}{1 \text{ lb.}} = \$2$ per pound.

Costs of Cheese

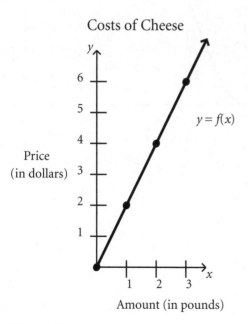

Figure 8.1

The slope is therefore the *rate* at which the price of cheese changes per pound. Because slope = $\frac{y\text{-change}}{x\text{-change}}$, a slope will always be a rate measured in y-units per x-unit.

For example, suppose a passenger on a bus writes down the exact time she passes each highway mile marker. She then sketches the graph shown in Figure 8.2 of the bus's position on the highway over time. The slope at any point on this graph will be measured in y-units per t-unit, or miles per hour. The steepness of the slope represents the speed of the bus.

1.

Figure 8.2

Practice

For each of the following four graphs, describe the rate that a slope of the curve represents.

2.

3.

Growth of a Baby

4.

Because the derivative of a function gives the slope of its tangent lines and the tangent line indicates how fast the function is increasing or decreasing, these practice problems show that the derivative of a function gives its *rate of change*.

An excellent example comes from position functions. A *position function* $s(t)$ states the position of an object along a straight line at any given time. The derivative $s'(t)$ states the rate at which that object's position is changing—that is, the *velocity* of the function. Thus, $s'(t) = v(t)$. The second derivative $s''(t) = v'(t)$ tells how fast the velocity is changing, or the *acceleration*. Thus, $s''(t) = v'(t) = a(t)$ where $s(t)$ is the position function, $v(t)$ is the velocity function, and $a(t)$ is the acceleration function.

Example

Suppose an object rolls along beside a tape measure so that after t seconds, it is next to the inch marked $s(t) = 4t^2 + 8t + 5$. Where is the object after 1 second? After 3 seconds? What is the velocity function? How fast is the object moving after 2 seconds? What is the acceleration function?

Solution

The position function $s(t) = 4t^2 + 8t + 5$ tells us where the object is. After 1 second, the object is next to the $s(1) = 17$-inch mark on the tape measure. After 3 seconds, the object is at the $s(3) = 65$-inch mark.

The velocity function is $v(t) = s'(t) = 8t + 8$. Thus, after 2 seconds, the object is moving at the rate of $v(2) = 24$ inches per second. Realize that this velocity of 24 inches per second is an *instantaneous velocity,* the speed just at a single moment. If a car's speedometer reads 60 miles per hour, this does not mean that it will drive for 60 miles or even for a full hour. The car might speed up, slow down, or stop. However, at that instant, the car is traveling at a rate that, if unchanged, will take it 60 miles in one hour. A derivative is always an instantaneous rate, telling you the slope at a particular point, but not making any promises about what will happen next.

The acceleration function is $a(t) = v'(t) = s''(t) = 8$. Because this is a constant, it tells us that the object increases in speed by 8 inches per second every second.

The most popular example of constant acceleration is gravity, which accelerates objects downward by $32\frac{\text{ft}}{\text{sec}}$ every second. Because of this, an object dropped with an initial velocity of b feet per second from a height of h feet above the ground will have (after t seconds) a height of $s(t) = -16t^2 + bt + h$ feet.

The starting time is $t = 0$, at which point the object is $s(0) = h$ feet off the ground, the correct initial height. The velocity function is $v(t) = s'(t) = -32t + b$. At the starting time $t = 0$, the velocity is $v(0) = b$, the desired initial velocity. The function $v(t) = -32t + b$ means that 32 feet per second are subtracted from the initial velocity b every second. The acceleration function is $a(t) = v'(t) = s''(t) = -32$. This is the desired constant acceleration.

Note: A negative velocity means the object is moving *backward* (or in the direction of decreasing y-value). The *speed* is the absolute value of the velocity.

Example

Suppose a brick is thrown straight upward with an initial speed of $10\frac{\text{ft}}{\text{sec}}$ from a 150-foot rooftop. What are its position, velocity, and acceleration functions?

Solution

Because the initial velocity is $b = 10\frac{\text{ft}}{\text{sec}}$ and the initial height is $h = 150$ feet, the position function is $s(t) = -16t^2 + 10t + 150$. The velocity function is $b(t) = s'(t) = -32t + 10$. The acceleration is $a(t) = -32$, a con-

stant 32 feet per second downward each second. The negative sign indicates that gravity is acting to decrease the height of the brick, pulling it downward.

Example

Suppose a rock is dropped from a 144-foot tall bridge. When will the rock hit the water? How fast will it be going then?

Solution

Because the rock is dropped, the initial velocity is $b = 0$. The initial height is $h = 144$. Thus, $s(t) = -16t^2 + 144$ gives the height function. The rock will hit the water (have a height of zero) when:

$$-16t^2 + 144 = 0$$

$$144 = 16t^2$$

$$t = \pm 3$$

And because -3 seconds doesn't make any sense, the rock will hit the water after 3 seconds.

The velocity function is $v(t) = s'(t) = -32t$; therefore, the rock will have a velocity of $v(3) = -96$ after 3 seconds. This means that it will be traveling at a rate of 96 feet per second downward when it hits the water.

Example

If $p(t) = \frac{t^2}{10} - 80t + 50{,}000$ gives the value, in thousands of dollars, of a start-up company after t days, then how fast is its value changing after 30 days? After 500 days?

Solution

The derivative $p'(t) = \dfrac{t}{5} - 80$ gives the rate of change in value, measured in thousands of dollars per day. After 30 days, $p'(30) = -74$, so the company will be losing value at a rate of $74,000 per day. After 500 days, $p'(500) = 20$, so the company will be gaining value at the instantaneous rate of $20,000 a day.

Practice

5. The height of a tree after t years is $h(t) = 30 - \dfrac{25}{t}$ feet when $t \geq 1$. How fast is the tree growing after 5 years?

6. The level of a river t days after a heavy rainstorm is $L(t) = -t^2 + 8t + 26$ feet. How fast is the river's level changing after 7 days?

7. When a company makes and sells x cars, its profit is $P(x) = \dfrac{x^3}{10} - 60x^2 + 9,000x$ dollars. How fast is its profit changing when the company makes 50 cars?

8. When a container is made x inches wide, it costs $C(x) = 0.8x^2 + \dfrac{24}{x}$ dollars to make. How is the cost changing when $x = 3$ inches?

9. An electron in a particle accelerator is $s(t) = t^3 + 2t^2 + 10t$ meters from the start after t seconds. Where is it after 3 seconds? How fast is it moving then? How fast is it accelerating then?

10. A brick is dropped from 64 feet above the ground. What is its position function? What is its velocity function? What is its acceleration? When will it hit the ground? How fast will it be traveling then?

11. A bullet is fired directly upward at 800 feet per second from the ground. How high is it when it stops rising and starts to fall?

12. A rock is thrown 10 feet per second down a 1,000-foot cliff. How far has it gone down in the first 4 seconds? How fast is it traveling then?

Derivatives of Sine and Cosine

Examining rates and slopes at various points can help us determine the derivative of $\sin(x)$. Look at the slopes at various points on its graph in Figure 8.2. It appears that the derivative function of $\sin(x)$ must oscillate between -1 and 1, and must go through the following points (see Figure 8.3). The function $\cos(x)$ is exactly such an oscillating function (see Figure 8.4). This suggests that $\dfrac{d}{dx}(\sin(x)) = \cos(x)$.

A similar study of the slopes of $\cos(x)$ would show that $\dfrac{d}{dx}(\cos(x)) = -\sin(x)$. The slopes of the cosine function are not the values of the sine function, but rather their exact negatives. These two results could be obtained using the limit definition of the derivative, but involve the use of trigonometric identities and certain limit formulas not covered in this book.

Examples

Differentiate the following

$$f(x) = 5\sin(x) + 4x^2$$

$$y = 2 + \cos(t)$$

⚠ **CAUTION**

$\dfrac{d}{dx}\sin(c) = 0$, not $\cos(c)$, when c is a constant. Similarly, $\dfrac{d}{dx}\cos(c) = 0$.

Figure 8.2

Figure 8.3

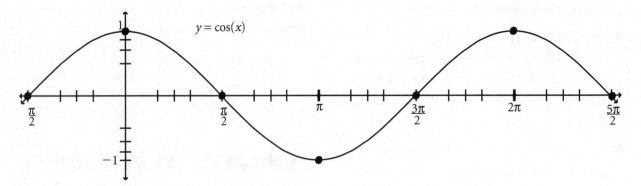

Figure 8.4

$$g(x) = \sin(x) - \cos(x) + \sin\left(\frac{\pi}{6}\right)$$

Solutions

$$f'(x) = 5\cos(x) + 8x$$

$$\frac{dy}{dt} = -\sin(t)$$

$$g'(x) = \cos(x) + \sin(x)$$

Practice

For questions 13 through 17, compute the derivative.

13. $y = 4x^5 + 10\cos(x) + 3$

14. $f(t) = 3\sin(t) + \dfrac{2}{t}$

15. $g(x) = 3x^{-2} + \dfrac{5}{\sqrt[3]{x}} - \pi\cos(x)$

16. $r(\theta) = \dfrac{1}{2}\sin(\theta) + \dfrac{1}{2}\cos(\theta)$

17. $h(x) = \cos(x) + \cos(5)$

18. Find the equation of the tangent line to
$f(x) = \sin(x) + \cos(x)$ at $x = \dfrac{\pi}{2}$.

Derivatives of the Exponential and Natural Logarithm Functions

The reason why the nicest exponential function is e^x where $e = 2.71828\ldots$ is because this makes for the following very nice derivative:

$$\frac{d}{dx}(e^x) = e^x$$

It is only with this exact base that the derivative of the exponential function is itself (see Figure 8.5).

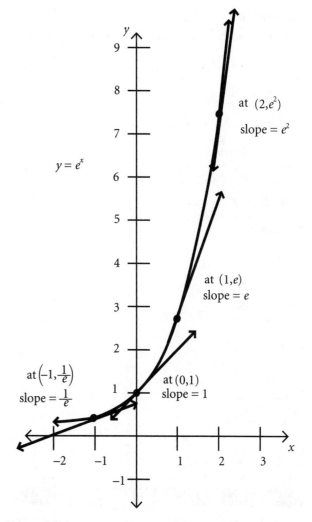

Figure 8.5

The derivative of its inverse function $\ln(x)$ is as follows:

$$\frac{d}{dx}(\ln(x)) = \frac{1}{x}$$

A proof of this formula is given in Lesson 11.

Examples

Differentiate the following.

$$f(x) = 4e^x$$

$$y = 10e^x + 10$$

$$g(t) = 3e^t + 2\ln(t)$$

$$y = 8\ln(u) - e^u + eu$$

$$y = e^3 + \ln(5)$$

Solutions

$$f'(x) = 4e^x$$

$$\frac{dy}{dx} = 10e^x$$

$$g'(t) = 3e^t + \frac{2}{t}$$

$$\frac{dy}{du} = \frac{8}{u} - e^u + e$$

$$\frac{dy}{dx} = 0$$

Practice

For questions 19 through 23, compute the derivative.

19. $f(x) = 1 + x + x^2 + x^3 + e^x$

20. $g(t) = 12\ln(t) + t^2 + 4$

21. $y = \cos(x) - 10e^x + 8x$

22. $h(x) = \sqrt{x} - 8\ln(x) + e^{-3}$

23. $k(x) = 3x^{\frac{5}{2}} + 5e^x + \ln(\pi)$

24. Find the second derivative of
$$f(x) = e^x + \ln(x).$$

25. Find the 100th derivative of $g(x) = 3e^x$.

26. What is the slope of the tangent line to
$$f(x) = \ln(x) \text{ at } x = 10?$$

⚠ **CAUTION**

e^x (e raised to the power x) and ex (the number e times x) are very different functions. Note that $\frac{d}{dx}(e^x) = e^x$, whereas $\frac{d}{dx}(ex) = e$.

⚠ **CAUTION**

$\frac{d}{dx}(e^c) = 0$ and $\frac{d}{dx}(\ln(c)) = 0$ whenever c is a positive constant.

9 ▶ THE PRODUCT AND QUOTIENT RULES

The Product Rule

When a function consists of parts that are added together, it is easy to take its derivative: Simply take the derivative of each part and add them together. We are inclined to try the same trick when the parts are multiplied together, but it does not work.

For example, we know that $\frac{d}{dx}(x^2) = 2x$ and $\frac{d}{dx}(x^3) = 3x^2$. The derivative of their product is

$\frac{d}{dx}(x^2 \cdot x^3) = \frac{d}{dx}(x^5) = 5x^4$. This shows that the derivative of a product is *not* the product of the derivatives:

$$5x^4 = \frac{d}{dx}(x^2 \cdot x^3) \neq \frac{d}{dx}(x^2) \cdot \frac{d}{dx}(x^3) = (2x) \cdot (3x^2) = 6x^3$$

Instead, we take the derivative of each part, multiply by the other part *left alone,* and add these results together:

$$\frac{d}{dx}(x^2 \cdot x^3) = \frac{d}{dx}(x^2) \cdot x^3 + \frac{d}{dx}(x^3) \cdot x^2 = (2x) \cdot x^3 + (3x^2) \cdot x^2 = 5x^4$$

This time, we *did* get the correct answer.

The Product Rule can be stated "the derivative of the first times the second, plus the derivative of the second times the first." It can be proven directly from the limit definition of the derivative, but only with a few tricks and a lot of algebra. Using Leibniz's notation, the Product Rule is stated as follows:

$$\frac{d}{dx}(f(x) \cdot g(x)) = f'(x) \cdot g(x) + g'(x) \cdot f(x)$$

Example
Differentiate $y = x^3\sin(x)$.

Solution
Here, the "first part" is x^3 and the "second part" is $\sin(x)$. Thus, by using the Product Rule, $\frac{d}{dx}(x^3\sin(x)) = \frac{d}{dx}(x^3) \cdot \sin(x) + \frac{d}{dx}(\sin(x)) \cdot x^3 = 3x^2\sin(x) + \cos(x) \cdot x^3$. This could be simplified as

$$\frac{dy}{dx} = x^2(3\sin(x) + x\cos(x)).$$

Example
Differentiate $f(x) = \ln(x) \cdot \cos(x)$.

Solution

$$f'(x) = \frac{d}{dx}(\ln(x)) \cdot \cos(x) + \frac{d}{dx}(\cos(x)) \cdot \ln(x)$$

$$= \frac{1}{x} \cdot \cos(x) - \sin(x) \cdot \ln(x)$$

Thus, the derivative is:

$$f'(x) = \frac{\cos(x)}{x} - \sin(x)\ln(x)$$

Example

Differentiate $g(x) = 5x^7 \cdot e^x$.

Solution

$$g'(x) = \frac{d}{dx}(5x^7) \cdot e^x + \frac{d}{dx}(e^x) \cdot 5x^7$$

$$= 35x^6 \cdot e^x + e^x \cdot 5x^7 = 5x^6e^x(7 + x)$$

Using the product rule with e^x can be a little bit confusing because there is no difference between the derivative of e^x and e^x "left alone." Still, if you write everything out, the correct answer should fall into place, even if it looks weird.

Example
Differentiate $y = \sqrt[3]{t} \cdot \ln(t)$.

Solution

$$\frac{dy}{dt} = \frac{d}{dt}\left(t^{\frac{1}{3}}\right) \cdot \ln(t) + \frac{d}{dt}\left(\ln(t)\right) \cdot t^{\frac{1}{3}}$$

$$= \frac{1}{3}t^{-\frac{2}{3}} \cdot \ln(t) + \left(\frac{1}{t}\right) \cdot t^{\frac{1}{3}}$$

$$= \frac{1}{3t^{\frac{2}{3}}} \cdot \ln(t) + \frac{1}{t^{\frac{2}{3}}}$$

$$= \frac{1}{t^{\frac{2}{3}}}\left[\frac{\ln(t)}{3} + 1\right]$$

Example
Differentiate $y = x^5\sin(x)\cos(x)$.

Solution
We'll use the Product Rule with x^5 as the first part and $\sin(x)\cos(x)$ as the second part. However, in taking

$$\frac{d}{dx}\left(\frac{f(x)}{g(x)}\right) = \frac{f'(x)g(x) - g'(x)f(x)}{(g(x))^2}$$

the derivative of $\sin(x)\cos(x)$, we'll have to use the Product Rule a second time. This can get messy, but it will be fine if everything is written down carefully.

$$\frac{dy}{dx} = \frac{d}{dx}(x^5) \cdot \sin(x)\cos(x) + \frac{d}{dx}(\sin(x)\cos(x)) \cdot x^5$$

$$= 4x^5 \sin(x)\cos(x)\left[\frac{d}{dx}(\sin(x)) \cdot \cos(x) + \frac{d}{dx}(\cos(x)) \cdot \sin(x)\right] \cdot x^5$$

$$= 4x^5 \sin(x)\cos(x)$$

$$\left[\cos(x) \cdot \cos(x) - \sin(x) \cdot \sin(x)\right] \cdot x^5$$

$$= 4x^5 \sin(x)\cos(x) + (\cos^2(x) - \sin^2(x)) \cdot x^5$$

Practice

For questions 1 through 12, compute the derivative.

1. $f(x) = x^2\cos(x)$

2. $y = 8t^3 e^t$

3. $y = \pi \sin(x)\cos(x)$

4. $g(x) = 3x^2\ln(x) - 5x^4 + 10$

5. $h(u) = \left(u^2 + 3u\right)e^u$

6. $k(x) = \sqrt{x}\cos(x) - \dfrac{\sin(x)}{\sqrt[4]{x}}$

7. $y = 8\ln(x)\sin(x) + \cos(x)$

8. $h(t) = \left(t^3 + 4\right)\left(\sin(t) - \cos(t)\right)$

9. $y = 5x^3 - x\ln(x)$

10. $f(x) = \sin^2(x) = \sin(x) \cdot \sin(x)$

11. $y = xe^x\sin(x)$. (Hint: $y = x \cdot \left(e^x \sin(x)\right)$.)

12. $g(x) = 3x^4\ln(x)\cos(x)$

13. What is the slope of the tangent line to $f(x) = x^2 e^x + x + 2$ at (0,2)?

14. Find the equation of the tangent line to $y = x\sin(x)$ at $x = \pi$.

The Quotient Rule

The Quotient Rule for functions where the parts are divided is slightly more complicated than the Product Rule. The Quotient Rule can be stated:

$$\frac{d}{dx}\left(\frac{f(x)}{g(x)}\right) = \frac{f'(x)g(x) - g'(x)f(x)}{(g(x))^2}$$

Just as with the Product Rule, each part is differentiated and multiplied by the other part. Here, however, they are subtracted, so it matters which one is differentiated first. It is important to start with the derivative of the top.

<table>
<tr><td>

⚠️ **CAUTION**

Just as with the Product Rule,

$$\frac{d}{dx}\left(\frac{f(x)}{g(x)}\right) \neq \frac{\frac{d}{dx}(f(x))}{\frac{d}{dx}(g(x))}.$$

</td></tr>
</table>

Example

Differentiate $y = \dfrac{x^5 - 3x^2 + 1}{\cos(x)}$.

Solution

Here, the top part is $x^5 - 3x^2 + 1$ and the bottom part is $\cos(x)$. Therefore, by the Quotient Rule:

$$\frac{dy}{dx} =$$

$$\frac{\frac{d}{dx}(x^5 - 3x^2 + 1)\cdot\cos(x) - \frac{d}{dx}(\cos(x))\cdot(x^5 - 3x^2 + 1)}{(\cos(x))^2}$$

$$= \frac{(5x^4 - 6x)\cdot\cos(x) - (-\sin(x))\cdot(x^5 - 3x^2 + 1)}{\cos^2(x)}$$

$$= \frac{(5x^4 - 6x)\cdot\cos(x) + \sin(x)\cdot(x^5 - 3x^2 + 1)}{\cos^2(x)}$$

Example

Differentiate $f(x) = \dfrac{x^3}{10x^2 - 1}$.

Solution

$$f'(x) = \frac{\frac{d}{dx}(x^3)\cdot(10x^2 - 1) - \frac{d}{dx}(10x^2 - 1)\cdot x^3}{(10x^2 - 1)^2}$$

$$= \frac{(3x^2)\cdot(10x^2 - 1)(20x)\cdot x^3}{(10x^2 - 1)}$$

$$= \frac{30x^4 - 3x^2 - 20x^4}{(10x^2 - 1)^2} = \frac{10x^4 - 3x^2}{(10x^2 - 1)^2}$$

Example

Differentiate $y = \dfrac{x^2\sin(x)}{\ln(x)}$.

Solution

Here, the Product Rule is necessary to differentiate the top.

$$\frac{dy}{dx} = \frac{\frac{d}{dx}(x^2\sin(x))\cdot\ln(x) - \frac{d}{dx}(\ln(x))\cdot x^2\sin(x)}{(\ln(x))^2}$$

$$= \frac{\left[\frac{d}{dx}(x^2)\cdot\sin(x) + \frac{d}{dx}(\sin(x))\cdot x^2\right]\cdot\ln(x) - \frac{1}{x}\cdot x^2\sin(x)}{(\ln(x))^2}$$

$$= \frac{\left[2x\cdot\sin(x) + \cos(x)\cdot x^2\right]\cdot\ln(x) - x\sin(x)}{(\ln(x))^2}$$

Example

Differentiate $y = \dfrac{\ln(t)}{t}$.

Solution

$$\frac{dy}{dt} = \frac{\frac{d}{dt}(\ln(t))\cdot t - \frac{d}{dt}(t)\cdot\ln(t)}{t^2}$$

$$= \frac{\frac{1}{t}\cdot t - 1\cdot\ln(t)}{t^2}$$

$$= \frac{1 - \ln(t)}{t^2}$$

Practice

For questions 15 through 26, compute the derivative.

15. $h(x) = \dfrac{x^3 + 10x - 7}{3x^2 + 5x + 2}$

16. $f(x) = \dfrac{x^2 - 1}{x^2 + 1}$

17. $f(x) = \dfrac{x + \ln(x)}{e^x - 1}$

18. $y = \dfrac{x^5}{\ln(x)}$

19. $y = \dfrac{4e^t + t}{t^3 + 2t + 1}$

20. $g(t) = \dfrac{t^3}{\pi \sin(t)}$

21. $y = \dfrac{1 - x\sqrt{x}}{1 + x\sqrt{x}}$

22. $g(u) = \dfrac{\sin(u)}{u^3 - 3u}$

23. $y = \dfrac{1}{\big(\sin(x) + \pi\big)\big(\cos(x) - \pi\big)}$

24. $h(t) = \dfrac{\ln(t) + t}{\sin^2(t)}$

25. $y = \dfrac{x\ln(x)}{e^x}$

26. $f(x) = \dfrac{x^2 e^x}{\cos(x)}$

27. Find the second derivative of $y = xe^x + ex$.

28. What is the slope of the tangent line to $f(x) = x^2 \ln(x)$ at $x = e$?

Derivatives of Trigonometric Functions

We can find the derivatives of the rest of the trigonometric functions using the Quotient Rule.

Example
Differentiate $y = \tan(x)$.

Solution

Use $\tan(x) = \dfrac{\sin(x)}{\cos(x)}$.

$$\frac{dy}{dx} = \frac{d}{dx}(\tan(x)) = \frac{d}{dx}\left(\frac{\sin(x)}{\cos(x)}\right)$$

Differentiate using the Quotient Rule.

$$\frac{dy}{dx} = \frac{\cos(x) \cdot \cos(x) - (-\sin(x)) \cdot \sin(x)}{\cos^2(x)}$$

Simplify.

$$\frac{dy}{dx} = \frac{\cos^2(x) + \sin^2(x)}{\cos^2(x)}$$

Use $\sin^2(x) + \cos^2(x) = 1$.

$$\frac{dy}{dx} = \frac{1}{\cos^2(x)}$$

Use $\sec(x) = \dfrac{1}{\cos(x)}$.

$$\frac{dy}{dx} = \sec^2(x)$$

Thus:

$$\frac{d}{dx}(\tan(x)) = \sec^2(x)$$

Example

Differentiate $y = \sec(x)$.

Solution

Use $\sec(x) = \dfrac{1}{\cos(x)}$.

$$\frac{dy}{dx} = \frac{d}{dx}(\sec(x)) = \frac{d}{dx}\left(\frac{1}{\cos(x)}\right)$$

Differentiate using the Quotient Rule.

$$\frac{dy}{dx} = \frac{0 \cdot \cos(x) - (-\sin(x)) \cdot 1}{\cos^2(x)}$$

Simplify.

$$\frac{dy}{dx} = \frac{\sin(x)}{\cos^2(x)} = \frac{1}{\cos(x)} \cdot \frac{\sin(x)}{\cos(x)}$$

Use $\sec(x) = \dfrac{1}{\cos(x)}$ and $\tan(x) = \dfrac{\sin(x)}{\cos(x)}$.

$$\frac{dy}{dx} = \sec(x)\tan(x)$$

Thus:

$$\frac{d}{dx}(\sec(x)) = \sec(x)\tan(x)$$

Practice

Differentiate the following.

29. $y = \csc(x)$

30. $y = \cot(x)$

31. $f(x) = x\tan(x)$

32. $g(x) = e^x \sec(x)$

33. $h(t) = e^t \ln(t)\tan(t)$

34. $j(x) = \dfrac{\sqrt[3]{x} + \sqrt[4]{x}}{\sec(x)}$

⚠ CAUTION

$e^t \cdot \ln(t)$ and $e^{\ln(t)}$ are NOT the same function. The first one is a product, whereas the second is a composition.

10 ▶ CHAIN RULE

We have learned how to compute derivatives of functions that are added, subtracted, multiplied, and divided. Next, we will learn how to compute the derivative of a composition of functions For example, it would be difficult to multiply out $f(x) = (x^3 + 10x + 4)^5$ just to take the derivative. Instead, notice that $f(x)$ looks like $g(x) = x^3 + 10x + 4$ put inside $h(x) = x^5$. Therefore, in terms of composition, $f(x) = (h \circ g)(x) = h(g(x))$.

The trick to differentiating composed functions is to take the derivative of the outermost layer first, while leaving the inner part alone, and then multiplying that by the derivative of the inside.

Using Leibniz's notation, the Chain Rule can be stated as follows:

$$\frac{d}{dx}(h(g(x))) = h'(g(x)) \cdot g'(x)$$

If this is confusing, think of the Chain Rule in the following way:

$$\frac{d}{dx}h(\text{something}) = h'(\text{something}) \cdot \frac{d}{dx}(\text{something})$$

$$\frac{d}{dx}(h(g(x))) = h'(g(x)) \cdot g'(x) \text{ or } \frac{d}{dx}(h(\text{something})) = h'(\text{something}) \cdot \frac{d}{dx}(\text{something})$$

The usual key to figuring out what is inside and what is outside is to watch the parentheses. Imagine that the parentheses form the layers of an onion, and that you must peel (differentiate) the outermost layers one at a time before reaching the inside.

Example

Differentiate $f(x) = (x^3 + 10x + 4)^5$.

Solution

Here, $f(x) = (\text{something})^5$ where the something $= x^3 + 10x + 4$. Because $\frac{d}{dx}(x^5) = 5x^4$, the Chain Rule gives:

$$f'(x) = 5(\text{something})^4 \cdot \frac{d}{dx}(\text{something})$$

$$= 5(x^3 + 10x + 4)^4 \cdot \frac{d}{dx}(x^3 + 10x + 4)$$

$$= 5(x^3 + 10x + 4)^4 \cdot (3x^2 + 10)$$

⚠ CAUTION

Make certain to not mix the derivatives of the layers to get and mistakenly say $f'(x) = 5(3x^2 + 10)^4$.

Example

Differentiate $g(x) = \sin(8x^4 + 3x^2 - 2x + 1)$.

Solution

Here, the function is essentially sin(something) where the "something" $= 8x^4 + 3x^2 - 2x + 1$. The derivative of sine is cosine, so the Chain Rule gives:

$$g'(x) = \cos(\text{something}) \cdot \frac{d}{dx}(\text{something})$$

$$g'(x) = \cos(8x^4 + 3x^2 - 2x + 1) \cdot$$

$$\frac{d}{dx}(8x^4 + 3x^2 - 2x + 1)$$

$$g'(x) = \cos(8x^4 + 3x^2 - 2x + 1) \cdot$$

$$(32x^3 + 6x - 2)$$

Example

Differentiate $y = \cos^3(x)$.

Solution

This is tricky because of the lack of parentheses. It might look like the "outside" function is cos(something), but it is actually $y = \cos^3(x) = (\cos(x))^3$. Thus, this function is really $(\text{something})^3$. So, the Chain Rule gives:

$$\frac{dy}{dx} = 3(\text{something})^2 \cdot \frac{d}{dx}(\text{something})$$

$$= 3(\cos)(x))^2 \cdot \frac{d}{dx}(\cos(x))$$

$$= 3(\cos)(x))^2 \cdot (-\sin(x))$$

$$= -3(\cos)^2 \cdot (x)\sin(x)$$

Example

Differentiate $y = \cos(x^3)$.

"SOMETHING" HINT

It is important that the "something" in the parentheses appear somewhere in the derivative, just as it does in the original function. If it doesn't appear, then a mistake has been made.

Solution

In this example, our function is cos(something). Because $\frac{d}{dx}(\cos(x)) = -\sin(x)$, the Chain Rule gives:

$$\frac{dy}{dx} = -\sin(\text{something}) \cdot \frac{d}{dx}(\text{something})$$

$$= -\sin(x^3) \cdot \frac{d}{dx}(x^3)$$

$$= -\sin(x^3) \cdot 3x^2$$

Example

Differentiate $h(x) = e^{5x}$.

Solution

$$h(x) = e^{(\text{something})}$$

so:

$$h'(x) = e^{(\text{something})} \cdot \frac{d}{dx}(\text{something})$$

$$= e^{5x} \cdot \frac{d}{dx}(5x) = e^{5x} \cdot 5 = 5e^{5x}$$

⚠ CAUTION

$\tan(\cos(x)) \neq \tan(x) \cdot \cos(x)$.

⚠ CAUTION

You CANNOT cancel the θ's to conclude that $\frac{\sin(2\theta)}{\theta} = \sin(2)$.

Practice

Differentiate the following.

1. $f(x) = (8x^3 + 7)^4$

2. $y = (x^2 + 8x + 9)^3$

3. $h(t) = (t^8 - 9t^3 + 3t + 2)^{10}$

4. $y = (u^5 - 3u^4 + 7)^{\frac{7}{2}}$

5. $g(x) = \sqrt{x^2 + 9x + 1}$

6. $y = \sqrt[3]{e^x + 1}$

7. $f(x) = \tan(\sqrt{x})$

8. $g(x) = \sqrt{\tan(x)}$

9. $y = \ln(3t + 5)$

10. $h(x) = \sin(\pi x)$

11. $y = (\ln(x))^5$

12. $f(x) = e^x + e^{2x} + e^{3x}$

13. $g(x) = \dfrac{e^x - e^{-x}}{2}$

14. $f(\theta) = \dfrac{\sin(2\theta)}{\theta}$

15. $y = xe^{2x}$

16. $f(x) = \sec(10x^2 + e^x)$

17. $y = e^{\left(x^2\right)} + x^{2e}$

18. $y = \ln(x\sin(x))$

19. $s(u) = (\sin(u) + \cos(u))^3$

20. $y = \tan(\cos(x))$

This rule is called the Chain Rule because it works in long succession when there are many layers to the function. It helps to write out the function using lots of parentheses, and then work patiently to take the derivative of each outermost layer.

Example

Differentiate $f(x) = \sin^7(e^{5x})$.

Solution

With all of its parentheses, this function is $f(x) = (\sin(e^{(5x)}))^7$. The outermost layer is "something to the seventh power," the second layer is "the sine of something," the third layer is "e raised to the something," and the last layer is $5x$. Thus:

$$f'(x) = 7(\sin(e^{(5x)}))^6 \cdot \frac{d}{dx}(\sin(e^{(5x)}))$$

$$= 7(\sin(e^{(5x)}))^6 \cdot \cos(e^{(5x)}) \cdot \frac{d}{dx}(e^{(5x)})$$

$$= 7(\sin(e^{(5x)}))^6 \cdot \cos(e^{(5x)}) \cdot e^{(5x)} \cdot \frac{d}{dx}(5x)$$

$$= 7(\sin(e^{(5x)}))^6 \cdot \cos(e^{(5x)}) \cdot e^{(5x)} \cdot 5$$

$$= 35e^{(5x)}\sin^6(e^{(5x)})\cos(e^{(5x)})$$

Example
Differentiate $y = \ln(x^3 + \tan(3x^2 + x))$.

Solution

$$\frac{dy}{dx} = \frac{1}{x^3 + \tan(3x^2 + x)} \cdot \frac{d}{dx}(x^3 + \tan(3x^2 + x))$$

$$= \frac{1}{x^3 + \tan(3x^2 + x)} \cdot \left(3x^2 + \sec^2(3x^2 + x) \cdot \frac{d}{dx}(3x^2 + x)\right)$$

$$= \frac{1}{x^3 + \tan(3x^2 + x)} \cdot (3x^2 + \sec^2(3x^2 + x) \cdot (6x + 1))$$

$$= \frac{3x^2 + \sec^2(3x^2 + x) \cdot (6x + 1)}{x^3 + \tan(3x^2 + x)}$$

Notice once again that every part except the outermost layer (the natural logarithm) appears somewhere in the derivative.

Practice

Differentiate the following.

21. $f(x) = \cos^3(8x)$

22. $y = (e^{9x^2 + 2x + 1})^4$

23. $g(t) = \ln(\tan(e^t + 1))$

24. $y = \sin(\sin(\sin(x)))$

25. $k(u) = \sec(\ln(8u^3))$

26. $h(x) = \cos^2\left(\dfrac{e^{4x^2} - 1}{1 - e^x}\right)$

LESSON 11 ▶ IMPLICIT DIFFERENTIATION

A common complaint about the Chain Rule is "I don't know where to stop!" For example, why do we use the Chain Rule for $f(x) = \sin(x^3)$ to get $f'(x) = \cos(x^3) \cdot 3x^2$, but not for $g(x) = \sin(x)$, which has $g'(x) = \cos(x)$? The honest answer is that we *could* use the Chain Rule as follows:

$$g'(x) = \cos(x) \cdot \frac{d}{dx}(x) = \cos(x) \cdot 1 = \cos(x)$$

$$f'(x) = \cos(x^3) \cdot \frac{d}{dx}(x^3) = \cos(x^3) \cdot 3x^2 \cdot \frac{d}{dx}(x) = \cos(x^3) \cdot 3x^2 \cdot 1 = \cos(x^3) \cdot 3x^2$$

When we get down to $\frac{d}{dx}(x) = 1$, we know we are done. The advantage to this way of thinking is that it explains what $\frac{dy}{dx}$ really means. This isn't merely a symbol that says "we took the derivative." This is the result of differentiating both sides of an equation.

Example

Differentiate $y = 4x^5 + e^x$.

Solution

Start with the equation.

$$y = 4x^5 + e^x$$

Differentiate both sides of the equation.

$$\frac{d}{dx}(y) = \frac{d}{dx}(4x^5 + e^x)$$

Use $\frac{d}{dx}(y) = \frac{dy}{dx}$.

$$\frac{dy}{dx} = 20x^4 \cdot \frac{d}{dx}(x) + e^x \cdot \frac{d}{dx}(x)$$

Simplify.

$$\frac{dy}{dx} = 20x^4 \cdot 1 + e^x \cdot 1 = 20x^4 + e^x$$

Now if $y = 4x^5 + e^x$, then there is a relationship between y and x. This relationship is given *explicitly* because we know exactly what y is in terms of x. However, if the variables x and y are all mixed up on both sides of the equals sign, then the relationship is said to be *implicit*. The relationship is implied, but it is up to us to figure out what the relationship between x and y is explicitly. For example, the equation of the unit circle is:

$$x^2 + y^2 = 1$$

There is a relationship between the values of x and y, because what y can be depends on the value of x. If $x = 0$, for instance, then y could be either 1 or -1. We could take the implicit description of y in $x^2 + y^2 = 1$ and make it explicit by solving for y:

$$y^2 = 1 - x^2$$

$$y = \pm\sqrt{1 - x^2}$$

Solving for y is not always possible, though. In fact, it rarely is. If our equation were $\ln(y) + \cos(y) = 3e^x - x^3$, then we would not be able to solve for y.

Fortunately, we can still find the slope of the tangent line, $\frac{dy}{dx}$, without having to solve the original equation for y. The trick is to use *implicit differentiation* by taking the derivative of both sides and making sure to include $\frac{d}{dx}(y) = \frac{dy}{dx}$ wherever the Chain Rule dictates.

Example

Find the slope of the tangent line to $x^2 + y^2 = 1$.

Solution

Start with the equation.

$$x^2 + y^2 = 1$$

Differentiate both sides.

$$\frac{d}{dx}(x^2 + y^2) = \frac{d}{dx}(1)$$

Use the Chain Rule everywhere.

$$2x \cdot \frac{d}{dx}(x) + 2y \cdot \frac{d}{dx}(y) = 0$$

Use $\frac{d}{dx}(x) = 1$ and $\frac{d}{dx}(y) = \frac{dy}{dx}$.

$$2x \cdot 1 + 2y \cdot \frac{dy}{dx} = 0$$

Solve for $\frac{dy}{dx}$.

$$\frac{dy}{dx} = \frac{-2x}{2y} = -\frac{x}{y}$$

It might make you uneasy to have $\frac{dy}{dx}$ given in terms of both x and y, but this is necessary. If we were asked, "What is the slope of the tangent line to $x^2 + y^2 = 1$ at $x = \frac{1}{2}$?" We would have to reply, "Which one?" There are *two* tangent lines when $x = \frac{1}{2}$! See Figure 11.1. If we want the slope of the tangent line at $\left(\frac{1}{2}, -\frac{\sqrt{3}}{2} \right)$, then

$$\frac{dy}{dx} = \frac{x}{y} = \frac{\frac{1}{2}}{-\frac{\sqrt{3}}{2}} = \frac{1}{\sqrt{3}} = \frac{\sqrt{3}}{3}.$$

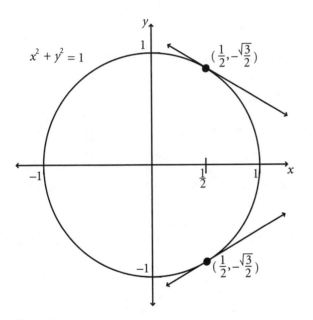

Figure 11.1

Example

Find $\frac{dy}{dx}$ when $\ln(y) + \cos(y) = 3e^x - x^3$.

Solution

Start with the equation.

$$\ln(y) + \cos(y) = 3e^x - x^3$$

> ⚠ **CAUTION**
>
> Don't forget to apply the Chain Rule when computing $\frac{d}{dx}(\cos(y))$ when y depends on x.

Differentiate both sides of the equation.

$$\frac{d}{dx}(\ln(y) + \cos(y)) = \frac{d}{dx}(3e^x - x^3)$$

Use the Chain Rule everywhere.

$$\frac{1}{y} \cdot \frac{d}{dx}(y) - \sin(y) \cdot \frac{d}{dx}(y) =$$
$$3e^x \cdot \frac{d}{dx}(x) - 3x^2 \cdot \frac{d}{dx}(x)$$

Use $\frac{d}{dx}(x) = 1$ and $\frac{d}{dx}(y) = \frac{dy}{dx}$.

$$\frac{1}{y} \cdot \frac{dy}{dx} - \sin(y) \cdot \frac{dy}{dx} = 3e^x - 3x^2$$

Factor out a $\frac{dy}{dx}$.

$$\left(\frac{1}{y} - \sin(y) \right) \frac{dy}{dx} = 3e^x - 3x^2$$

Solve for $\frac{dy}{dx}$.

$$\frac{dy}{dx} = \frac{3e^x - 3x^2}{\frac{1}{y} - \sin(y)}$$

To get rid of the fraction-in-a-fraction, we can multiply the top and bottom by the denominator y that we want to eliminate:

$$\frac{dy}{dx} = \frac{3e^x - 3x^2}{\frac{1}{y} - \sin(y)}$$
$$= \left(\frac{3e^x - 3x^2}{\frac{1}{y} - \sin(y)} \right) \cdot \left(\frac{y}{y} \right)$$
$$= \frac{3ye^x - 3x^2 y}{1 - y\sin(y)}$$

Example

Find the slope of the tangent line to $y^2\ln(x) = y + 5$ at $(1, -5)$.

Solution

Start with the equation.

$y^2\ln(x) = y + 5$

Differentiate both sides of the equation.

$\dfrac{d}{dx}(y^2\ln(x)) = \dfrac{d}{dx}(y + 5)$

Use the product rule on $y^2\ln(x)$.

$2y \cdot \dfrac{d}{dx}(y) \cdot \ln(x) + \dfrac{1}{x} \cdot \dfrac{d}{dx}(x) \cdot y^2 = \dfrac{d}{dx}(y) + 0$

Use $\dfrac{d}{dx}(x) = 1$ and $\dfrac{d}{dx}(y) = \dfrac{dy}{dx}$.

$2y \cdot \dfrac{dy}{dx} \cdot \ln(x) + \dfrac{1}{x} \cdot y^2 = \dfrac{dy}{dx}$

Plug in $x = 1$ and $y = -5$.

$2(-5) \cdot \dfrac{dy}{dx} \cdot \underbrace{\ln(1)}_{=0} + \dfrac{1}{1} \cdot (-5)^2 = \dfrac{dy}{dx}$

Use $\ln(1) = 0$.

$25 = \dfrac{dy}{dx}$

Thus, the slope of the tangent line at $(1, -5)$ is 25.

Example

Find $\dfrac{dy}{dx}$ when $\cos(x \cdot \sin(y))$.

Solution

Start with the equation.

$\tan(y) = \cos\big(x \cdot \sin(y)\big)$

Differentiate both sides of the equation.

$\dfrac{d}{dx}\big(\tan(y)\big) = \dfrac{d}{dx}\big(\cos\big(x \cdot \sin(y)\big)\big)$

Compute the derivatives on both sides.

$\sec^2(y) \cdot \dfrac{d}{dx}(y)$

$= -\sin\big(x \cdot \sin(y)\big) \cdot \left(\sin(y) + \left(\cos(y) \cdot \dfrac{d}{dx}(y)\right) \cdot x\right)$

Use $\dfrac{d}{dx}(x) = 1$ and $\dfrac{d}{dx}(y) = \dfrac{dy}{dx}$.

$\sec^2(y) \cdot \dfrac{dy}{dx} = -\sin\big(x \cdot \sin(y)\big) \cdot \left(\sin(y) + \left(\cos(y) \cdot \dfrac{dy}{dx}\right) \cdot x\right)$

Simplify the right-hand side.

$\sec^2(y) \cdot \dfrac{dy}{dx}$

$= -\sin\big(x \cdot \sin(y)\big) \cdot \sin(y) - \sin\big(x \cdot \sin(y)\big) \cdot \cos(y) \cdot x \cdot \dfrac{dy}{dx}$

Bring both instances of $\dfrac{dy}{dx}$ to the same side.

$\sec^2(y) \cdot \dfrac{dy}{dx} + \sin\big(x \cdot \sin(y)\big) \cdot \cos(y) \cdot x \cdot \dfrac{dy}{dx}$

$= -\sin\big(x \cdot \sin(y)\big) \cdot \sin(y)$

Factor out a $\dfrac{dy}{dx}$.

$\left[\sec^2(y) + \sin\big(x \cdot \sin(y)\big) \cdot \cos(y) \cdot x\right]\dfrac{dy}{dx}$

$= -\sin\big(x \cdot \sin(y)\big) \cdot \sin(y)$

Solve for $\dfrac{dy}{dx}$.

$$\frac{dy}{dx} = \frac{-\sin\big(x \cdot \sin(y)\big) \cdot \sin(y)}{\sec^2(y) + \sin\big(x \cdot \sin(y)\big) \cdot \cos(y) \cdot x}$$

Example

Use implicit differentiation and the fact that $\dfrac{d}{dx}(e^x) = e^x$ to prove that $\dfrac{d}{dx}(\ln(x)) = \dfrac{1}{x}$.

Solution

If $y = \ln(x)$, then the derivative of $\ln(x)$ is $\dfrac{dy}{dx}$.

$$y = \ln(x)$$

Raise both sides as powers of e.

$$e^y = e^{\ln(x)}$$

Since $\ln(x)$ and e^x are inverses, $e^{\ln(x)} = x$.

$$e^y = x$$

Differentiate both sides.

$$\frac{d}{dx}(e^y) = \frac{d}{dx}(x)$$

Use the Chain Rule.

$$e^y \cdot \frac{dy}{dx} = 1$$

Solve for $\dfrac{dy}{dx}$.

$$\frac{dy}{dx} = \frac{1}{e^y}$$

Use $e^y = e^{\ln(x)} = x$.

$$\frac{dy}{dx} = \frac{1}{x}$$

So, $\dfrac{d}{dx}\big(\ln(x)\big) = \dfrac{1}{x}$.

Practice

For questions 1 through 14, compute $\dfrac{dy}{dx}$.

1. $(y + 1)^3 = x^4 - 8x$

2. $y^3 + y = \sin(x)$

3. $\sin(y) = 4x + 7$

4. $y - \sqrt{y} = \ln(x)$

5. $y^2 + x = 3x^4 + 8y$

6. $e^x + e^{2x} = e^y + e^{2y}$

7. $\tan(y) = \cos(x)$

8. $y = \sqrt{x + y}$

9. $\sin(x) - \sin(y) = \sin(x - y)$

10. $y - \ln(x) = y^3 - y^2 - y - 1$

11. $(y + x^2)^4 = 10x$

12. $x^2 y = y^4 x^4$

13. $\dfrac{x}{y} + xy = x + y$

14. $\sec(y) + 9y = x^3\cos(y)$

15. Find the tangent line slope of
$y^3 + x^2 = y^2 - 5y + 14$ at $(-3, 1)$.

16. Find the tangent line slope of
$x^3 + y^3 = 3y - x$ at $(1, -2)$.

⚠ CAUTION

$\sqrt{a+b} \neq \sqrt{a} + \sqrt{b}$. To see this, let $a = 16$ and $b = 9$. The left-hand side is $\sqrt{25} = \sqrt{5}$, but the right-hand side is $\sqrt{16} + \sqrt{9} = 4 + 3 = 7$.

> ⚠ **CAUTION**
>
> $\sin(A - B) \neq \sin(A) - \sin(B)$. To see this, let $A = \pi$ and $B = \left(\dfrac{\pi}{2}\right)$. The left-hand side is $\sin\left(\dfrac{\pi}{2}\right) = 1$, but the right-hand side is $\sin(\pi) - \sin\left(\dfrac{\pi}{2}\right) = 0 - 1 = -1$.

17. Find the slope of the tangent line to

$\ln(3y - 5) + x = y^2$ at $(4,2)$.

18. Find the slope of the tangent line at $\left(\dfrac{\sqrt{2}}{2}, \dfrac{\pi}{4}\right)$

on the graph of $\cos\left(\dfrac{\pi}{2\sqrt{2}} \sin(y)\right) = x$.

19. Find the equation of the tangent line to

$\sin(y) = x$ at the point $\left(\dfrac{1}{2}, \dfrac{\pi}{6}\right)$.

20. Find the equation of the tangent line to

$\ln\left(e^x + e^y\right) = \ln(2) - y$ at $(0,0)$.

12 ▶ RELATED RATES

Sometimes, both variables x and y depend on a third variable t. An equation relating x and y is often able to be determined geometrically. Once you have gotten the hang of implicit differentiation, it should not be difficult to take the derivative of both sides with respect to the variable t. This enables us to see how x and y vary with respect to time t. The only difference is that $\frac{d}{dt}(x) = \frac{dx}{dt}$, $\frac{d}{dt}(y) = \frac{dy}{dt}$, and so on. Only $\frac{d}{dt}(t) = 1$ can be simplified.

Example
Assume x and y depend on some variable t. Differentiate $y^2 + \cos(x) = 4x^2y$ with respect to t.

Solution
Start with the equation.

$$y^2 + \cos(x) = 4x^2y$$

Us the Chain Rule to differentiate both sides with respect to t.

$$\frac{d}{dt}(y^2 + \cos(x)) = \frac{d}{dt}(4x^2y)$$

$$2y \cdot \frac{d}{dt}(y) - \sin(x) \cdot \frac{d}{dt}(x) =$$

$$8x \cdot \frac{d}{dt}(x) \cdot y + \frac{d}{dt}(y) \cdot 4x^2$$

Use $\frac{d}{dt}(x) = \frac{dx}{dt}$ and $\frac{d}{dt}(y) = \frac{dy}{dt}$.

$$2y \cdot \frac{dy}{dt} - \sin(x) \cdot \frac{dx}{dt} = 8xy \cdot \frac{dx}{dt} + \frac{dy}{dt} \cdot 4x^2$$

Example

Assume x and y depend on some variable t. Differentiate $e^x + y = y^3 + \sqrt{x}$ with respect to t.

Solution

Start with the equation.

$$e^x + y = y^3 + \sqrt{x}$$

Differentiate both sides with respect to t.

$$\frac{d}{dt}(e^x + y) = \frac{d}{dt}(y^3 + \sqrt{x})$$

Use the Chain Rule everywhere.

$$e^x \cdot \frac{d}{dt}(x) + \frac{d}{dt}(y) = 3y^2 \cdot \frac{d}{dt}(y) + \frac{1}{2\sqrt{x}} \cdot \frac{d}{dt}(x)$$

Use $\frac{d}{dt}(x) = \frac{dx}{dt}$ and $\frac{d}{dt}(y) = \frac{dy}{dt}$.

$$e^x \cdot \frac{dx}{dt} + \frac{dy}{dt} = 3y^2 \cdot \frac{dy}{dt} + \frac{1}{2\sqrt{x}} \cdot \frac{dx}{dt}$$

The variables need not be x and y, and the variable upon which they depend need not be t.

Example

Assume A, B, and C depend on some variable r. Differentiate $3A + 4B^2 = \frac{A}{C} + \pi$.

Solution

$$\frac{d}{dr}(3A + 4B^2) = \frac{d}{dr}\left(\frac{A}{C} + \pi\right)$$

$$3 \cdot \frac{dA}{dr} + 8B \cdot \frac{dB}{dr} = \frac{\frac{dA}{dr} \cdot C - \frac{dC}{dr} \cdot A}{C^2}$$

⚠ CAUTION

$\frac{d}{dr}(\pi) = 0$ because π is a constant.

Practice

Assume all variables depend on t. Differentiate with respect to t.

1. $y = (x^3 + x - 1)^5$

2. $y^4 - 3x^2 = \cos(y)$

3. $\sqrt{x} + \sqrt{y} = 10x^3 - 7x$

4. $\ln(y) + e^x = x^2y^2$

5. $z = \frac{2}{5}x^2 + \frac{2}{5}y^2 + \frac{3}{5x}$

6. $A^2 + B^2 = C^2$

7. $V = \frac{4}{3}\pi r^3$

8. $A = 4\pi r^2$

9. $C = 2\pi r$

10. $A = \frac{1}{2}bh$

11. $S = 6e^2$

12. $D = \sqrt{x^2 + y^2}$

Just as $\dfrac{dy}{dx} = \dfrac{y\text{-change}}{x\text{-change}}$ is a rate, so are $\dfrac{dx}{dt}$, $\dfrac{dy}{dt}$, $\dfrac{dA}{dt}$, and so on. Because t typically represents time, $\dfrac{dy}{dt} = \dfrac{y\text{-change}}{t\text{-change}}$ represents how fast y is changing over time. Thus, if A is a variable that represents an area, $\dfrac{dA}{dt}$ represents how fast that area is increasing or decreasing over time.

Differentiating an equation with respect to t results in a new equation, which shows how the rates of change of the variables are related. For example, the area A and radius r of a circle are related by:

$$A = \pi r^2$$

Differentiating both sides with respect to t gives:

$$\frac{dA}{dt} = 2\pi r \cdot \frac{dr}{dt}$$

If a circle is growing in size, this equation details how the rate at which the radius is changing, $\dfrac{dr}{dt}$, relates to the rate at which the area is growing, $\dfrac{dA}{dt}$.

Example

A rock thrown into a pond makes a circular ripple that travels at 4 feet per second. How fast is the area of the circle increasing when the circle has a radius of 12 feet?

Solution

We know that for circles $A = \pi r^2$, so that, $\dfrac{dA}{dt} = 2\pi r \cdot \dfrac{dr}{dt}$. And we know that the radius is increasing at the rate of $\dfrac{dr}{dt} = 4$ feet per second, so when the radius is $r = 12$ feet, the area is increasing at:

$$\frac{dA}{dt} = 2\pi\,(12 \text{ feet}) \cdot 4\frac{\text{feet}}{\text{second}}$$

$$= 96\pi\frac{\text{ft}^2}{\text{sec}}$$

$$= 96\pi \approx 301.6 \text{ square feet per second}$$

Example

A spherical balloon is inflated with 40 cubic inches of air every second. When the radius is 12 inches, how fast is the radius of the balloon increasing? (Hint: The volume of a sphere with radius r is $V = \frac{4}{3}\pi r^3$.)

Solution

We know that the volume of the balloon is increasing at the rate of $\dfrac{dV}{dt} = 40\dfrac{\text{in}^3}{\text{sec}}$. We want to know the value of $\dfrac{dr}{dt}$ when $r = 12$ inches. Differentiating $V = \dfrac{4}{3}\pi r^3$ with respect to t gives:

$$\frac{dV}{dt} = 4\pi r^2 \cdot \frac{dr}{dt}$$

When we plug in $\dfrac{dV}{dt} = 40\dfrac{\text{in}^3}{\text{sec}}$ and $r = 12$ in, we get:

$$40\frac{\text{in}^3}{\text{sec}} = 4\pi\,(12\,\text{in})^2 \cdot \frac{dr}{dt}$$

$$\frac{dr}{dt} = \frac{40 \text{ in.}}{4\pi \cdot 144 \text{ sec}} = \frac{5 \text{ in.}}{72\pi\,\text{sec}}$$

The radius of the balloon is increasing at the very slow rate of $\dfrac{5}{72\pi} \approx 0.022$ inches per second.

Example

Suppose the base of a triangle is increasing at a rate of 8 feet per minute while the height is decreasing by 1 foot every minute. How fast is the triangle's area changing when the height is 5 feet and the base is 20 feet?

Solution

If we represent the length of the base by b, the height of the triangle as h, and the area of the triangle as A, then the formula that relates them all is $A = \frac{1}{2}bh$. The base is increasing at $\frac{db}{dt} = 8\frac{ft}{min}$ and the height is changing at $\frac{dh}{dt} = -1\frac{ft}{min}$. The -1 implies that 1 foot is subtracted from the height every minute, that is, the height is decreasing. We are trying to find $\frac{dA}{dt}$, which is the rate of change in area. When we differentiate the formula $A = \frac{1}{2}bh$ with respect to t, we get:

$$\frac{dA}{dt} = \frac{1}{2} \cdot \frac{db}{dt} \cdot h + \frac{dh}{dt} \cdot \frac{1}{2}b$$

When we plug in all of our information, including the $h = 5$ feet and $b = 20$ feet, we get:

$$\frac{dA}{dt} = \frac{1}{2} \cdot \left(8\frac{ft}{min}\right) \cdot \left(5\ ft\right) + \left(-1\frac{ft}{min}\right) \cdot \frac{1}{2} \cdot \left(20\ ft\right)$$

$$= 20\frac{ft^2}{min} - 10\frac{ft^2}{min} = 10\frac{ft^2}{min}$$

Thus, at the exact instant when the height is 5 feet and the base is 20 feet, the area of the triangle is increasing at a rate of 10 square feet every minute.

Example

A 20-foot ladder slides down a wall at the rate of 2 feet per minute (see Figure 12.1). How fast is it sliding along the ground when the ladder is 16 feet up the wall?

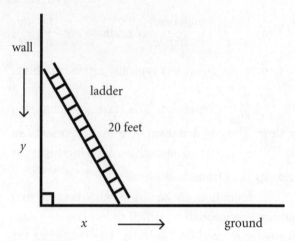

Figure 12.1

Solution

Here, $\frac{dy}{dt} = -2\frac{ft}{min}$ because the ladder is sliding down the wall at 2 feet per minute. We want to know $\frac{dx}{dt}$, the rate at which the bottom of the ladder is moving away from the wall. The equation to use is the Pythagorean theorem.

$$x^2 + y^2 = 20^2$$

$$\frac{d}{dt}(x^2 + y^2) = \frac{d}{dt}(20^2)$$

$$2x \cdot \frac{dx}{dt} + 2y \cdot \frac{dy}{dt} = 0$$

If we plug in $y = 16$ ft and $\frac{dy}{dt} = -2$ ft/min, we get:

$$2x \cdot \frac{dx}{dt} + 2 \cdot \left(16\ ft\right) \cdot \left(-2\frac{ft}{min}\right) = 0$$

We still need to know what x is at the particular instant that $y = 16$, and for this, we go back to the Pythagorean theorem.

$$x^2 + (16)^2 = (20)^2$$

$$x^2 = 144, \text{ so } x = \pm 12$$

It is important to use variables for all of the values that are changing. Only *after* all derivatives have been computed can they be replaced by numbers.

Using $x = 12$ (a negative length here makes no sense), we get:

$$2 \cdot (12 \text{ ft}) \cdot \frac{dx}{dt} + 2 \cdot (16 \text{ ft}) \cdot \left(-2 \frac{\text{ft}}{\text{min}}\right) = 0$$

$$\frac{dx}{dt} = \frac{8}{3} \frac{\text{ft}}{\text{min}}$$

At the moment that $y = 16$ ft, the ladder is sliding along the ground at $\frac{8}{3}$ feet per minute.

In the previous example, it was okay to say that the hypotenuse was 20 because the length of the ladder didn't change. However, if we replace y with 16 in the equation *before* differentiating, we would have implied that the height was fixed at 16 feet. Because the height *does* change, it needs to be written as a variable, y. In general, any quantity that varies needs to be represented with a variable. Only after all derivatives have been computed can the information for the given instant, like $y = 16$, be substituted.

Practice

13. Suppose $y^2 + 3y = 6 - 4x^3$ and $\frac{dy}{dt} = 5$. What is $\frac{dx}{dt}$ when $x = -1$ and $y = 2$?

14. Suppose $xy^2 = x^2 + 3$. What is $\frac{dy}{dt}$ when $\frac{dx}{dt} = 8$, $x = 3$, and $y = -2$?

15. Let $K + e^L = L + I^2$. If $\frac{dL}{dt} = 5$ and $\frac{dI}{dt} = 4$, what is $\frac{dK}{dt}$ when $L = 0$ and $I = 3$?

16. Suppose $A^3 = B^2 + 4C^2$, $\frac{dA}{dt} = 8$, and $\frac{dC}{dt} = -2$. What is $\frac{dB}{dt}$ when $A = 2$, $B = 2$, and $C = 1$?

17. Suppose $A = I^2 + 6R$. If I increases by 4 feet per minute and R increases by 2 square feet every minute, how fast is A changing when $I = 20$?

18. Suppose $K^3 = \frac{1}{R^2} + 11$. Every hour, K decreases by 2. How fast is R changing when $K = 3$ and $R = \frac{1}{4}$?

19. The height of a triangle decreases by 2 feet every minute while its base shrinks by 6 feet every minute. How fast is the area changing when the height is 15 feet and the base is 20 feet?

20. The surface area of a sphere with radius r is $A = 4\pi r^2$. If the radius is decreasing by 2 inches every hour, how fast is the surface area shrinking when the radius is 20 inches?

21. A circle increases in area by 20 square feet every hour. How fast is the radius increasing when the radius is 4 feet?

22. The volume of a cube grows by 1,200 cubic inches every minute. How fast is each side growing when each side is 10 inches?

23. The surface area of a cube is decreasing at a rate of 2 square inches per second. How fast is an edge shrinking at the instant when each side is 40 inches? (Hint: The surface area of a cube with edge e is $S = 6e^2$.)

24. The height of a triangle grows by 5 inches each hour. The area is increasing by 100 square inches each hour. How fast is the base of the triangle increasing when the height is 20 inches and the base is 12 inches?

25. One end of a 10-foot long board is lifted straight off the ground at 1 foot per second (see Figure 12.2). How fast will the other end drag along the ground after 6 seconds?

26. A kite is 100 feet off the ground and moving horizontally at 13 feet per second (see Figure 12.3). How quickly is the string being let out when the string is 260 feet long?

Figure 12.3

Figure 12.2

13 ▶ LIMITS AT INFINITY

This lesson will serve as a preparation for the graphing in the next lesson. Here, we will work on ways to identify asymptotes from the formula of a *rational function*, a quotient of two polynomials.

We've encountered *vertical asymptotes* informally in Lesson 5. They are easy to recognize for rational functions because they occur at precisely those *x*-values at which the denominator equals zero *and* the numerator does NOT equal zero. If *both* top and bottom are zero when evaluated at an x-value, you get a small unshaded circle on its graph at that point. For example, $f(x) = \dfrac{(3x + 2)(x - 1)}{(x + 3)(x - 4)}$ has vertical asymptotes at $x = -3$ and $x = 4$.

Horizontal asymptotes take a bit more work to identify. The graph will flatten out like a horizontal line if large values of *x* all have essentially the same *y*-value.

In the graph of $y = f(x)$, in Figure 13.1 for example, if *x* is bigger than 5, then *y* will be very close to *y* = 1. Similarly, if *x* is a large negative number, the corresponding *y*-value will be close to zero. Horizontal asymptotes are related to the limits as *x* gets really big. For $f(x)$ given in the graph:

$$\lim_{x \to \infty} f(x) = 1 \text{ and } \lim_{x \to -\infty} f(x) = 0$$

In such case, we say that $y = 1$ and $y = 0$ are horizontal asymptotes of *f*.

Notice that the graph of $y = f(x)$ crosses *both* horizontal asymptotes. Vertical asymptotes cannot be crossed because they are, by definition, not in the domain. Horizontal asymptotes *can* be crossed, as illustrated in this example. Think of "asymptote" as meaning "flattens out like a straight line" and not "a line not to be crossed."

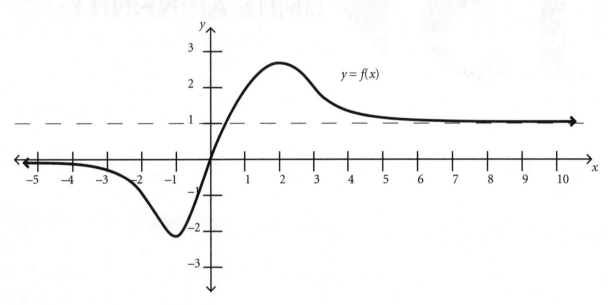

Figure 13.1

These *limits at infinity* (and negative infinity) identify what the ends of the graph do. For example, if $\lim_{x \to \infty} g(x) = 3$, then the graph of $y = g(x)$ will typically look something like that in Figure 13.2. If $\lim_{x \to -\infty} h(x) = \infty$, then the graph of $y = h(x)$ will look like that in Figure 13.3.

Notice that the infinite limits say only what happens way off to the left and to the right. Other calculations must be done to know what happens in the middle of the graph.

The general trick to evaluating an infinite limit is to focus on the most powerful part of the function. Take $\lim_{x \to \infty} (2x^3 - 100x^2 - 10x - 5{,}000)$ for example.

There are several terms being added in this function. However, the most powerful part is the term $2x^3$. When x gets big enough, like when $x = 1{,}000{,}000$, then

$$2x^3 - 100x^2 - 10x - 5{,}000$$

$$= 2{,}000{,}000{,}000{,}000{,}000{,}000 -$$

$$100{,}000{,}000{,}000{,}000 - 10{,}000{,}000 - 5{,}000$$

$$= 1{,}999{,}899{,}999{,}989{,}995{,}000$$

This clearly rounds to $2{,}000{,}000{,}000{,}000{,}000{,}000$, which is the value of $2x^3$ at $x = 1{,}000{,}000$. It is in this sense that $2x^3$ is called the most powerful part of the function. As x gets big, $2x^3$ is the only part that counts.

The rules for Infinite Limits of Rational Functions are as follows:

- If the numerator is more powerful, the limit goes to ∞ or −∞.
- If the denominator is more powerful, the limit goes to 0.
- If the numerator and denominator are evenly matched, the limit is formed by the coefficients of the most powerful parts.

OR

Figure 13.2

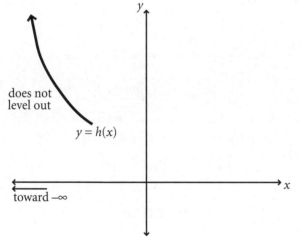

Figure 13.3

$$\lim_{x \to \infty}(2x^3 - 100x^2 - 10x - 5,000) = \lim_{x \to \infty} 2x^3 = \infty$$

As x gets huge, x^3 is clearly even larger, and $2x^3$ is twice that. Thus, as x goes to infinity, so does $2x^3$. Basically, the higher the exponent of x, the more powerful it is. With that in mind, the rules for infinite limits of rational functions are fairly simple:

- If the numerator is more powerful, the limit goes to ∞ or − ∞.
- If the denominator is more powerful, the limit goes to 0.
- If the numerator and denominator are evenly matched, the limit is formed by the coefficients of the most powerful parts.

The whole concept of "going to infinity" might be a bit confusing. This really means "going toward infinity," because infinity is not reachable. Just know that "going to infinity" means that we see what happens when we plug really large numbers into the function, and that "going to negative infinity" means that we see what happens when we plug really large *negative* numbers into the function.

Example

Evaluate $\lim\limits_{x \to \infty} \dfrac{1 - x^2}{x^3 + 3x + 2}$.

Solution

The most powerful part of the numerator is $-x^2$, and in the denominator is x^3. Thus:

$$\lim_{x \to \infty} \frac{1 - x^2}{x^3 + 3x + 2} = \lim_{x \to \infty} \frac{-x^2}{x^3} = \lim_{x \to \infty} -\frac{1}{x} = 0$$

This limit is zero because the numerator is overpowered by the denominator. Also, as x gets really big, $\dfrac{1}{x}$ gets really close to zero. For example, when $x = 1,000$, then $\dfrac{1}{x} = \dfrac{1}{1,000} = 0.001$.

Example

Evaluate $\lim\limits_{x \to -\infty} \dfrac{3x^2 + 2x - 5}{(1 - 2x)(1 + 2x)}$.

Solution

Here, the numerator and denominator are evenly matched, with each having x^2 as its highest power of x.

$$\lim_{x \to -\infty} \frac{3x^2 + 2x - 5}{(1 - 2x)(1 + 2x)} = \lim_{x \to -\infty} \frac{3x^2 + 2x - 5}{1 - 4x^2}$$

$$= \lim_{x \to -\infty} \frac{3x^2}{-4x^2}$$

$$= \lim_{x \to -\infty} \frac{3}{-4} = -\frac{3}{4}$$

⚠ CAUTION

Make certain to fully expand the polynomials in the top and bottom of a rational function *before* identifying the dominating terms in each.

The limit is formed by the coefficients of the most powerful parts: 3 in the numerator and -4 in the denominator.

Example

Evaluate $\lim\limits_{x \to \infty} \dfrac{5x^{10} - 4x^5 + 7}{1 - x^2}$.

Solution

Here,

$$\lim_{x \to \infty} \frac{5x^{10} - 4x^5 + 7}{1 - x^2} = \lim_{x \to \infty} \frac{5x^{10}}{-x^2}$$

$$= \lim_{x \to \infty} -5x^8 = -\infty$$

As x goes to infinity, x^8 also gets really large, but the negative in the -5 reverses this and makes $-5x^8$ approach negative infinity.

Practice

Evaluate the following infinite limits.

1. $\lim\limits_{x \to \infty} \dfrac{5x^3 + 10x^2 - 2}{8x^4 + 1}$

2. $\lim\limits_{x \to -\infty} \dfrac{4x^3 + 10x^2 + 3x}{5x^3 + 8x - 1}$

3. $\displaystyle\lim_{x\to\infty}\frac{5x+2}{2x-1}$

4. $\displaystyle\lim_{x\to\infty}\frac{10x^3-3x-100}{2x+5}$

5. $\displaystyle\lim_{t\to-\infty}\frac{t+1}{t(t+4)(t-1)}$

6. $\displaystyle\lim_{t\to\infty}\frac{8t^4-3t^3+11}{(1-3t^2)(1+3t^2)}$

7. $\displaystyle\lim_{x\to\infty}\frac{5x^2(x+2)}{1-x^2}$

8. $\displaystyle\lim_{x\to-\infty}\frac{x^4+3x^2-8x+4}{x^2+2x+1}$

9. $\displaystyle\lim_{x\to-\infty}\frac{x^2-1}{x^2+1}$

10. $\displaystyle\lim_{x\to\infty}\frac{t+20,000}{t^2-6,000,000}$

The infinite limits of e^x and $\ln(x)$ can be seen from their graphs in Figure 13.4.

$$\lim_{x\to\infty}e^x=\infty \qquad \lim_{x\to-\infty}e^x=0 \qquad \lim_{x\to\infty}\ln(x)=\infty$$

In general, as x goes to infinity, e^x is more powerful than x raised to any number. The natural logarithm, however, goes to infinity slower than any power of x. It may look as though $y=\ln(x)$ is beginning to level out toward a horizontal asymptote, but actually, it will eventually surpass any height as it slowly goes up to infinity.

In more complicated situations, we use L'Hôpital's rule. This states that if the numerator and denominator both go to infinity (positive or negative), then the limit remains the same after taking the derivative of the top and the bottom.

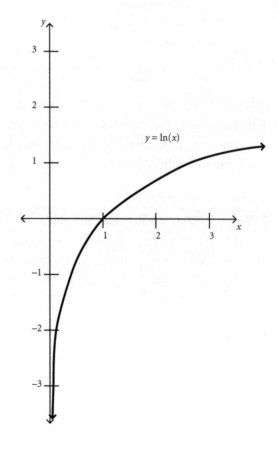

Figure 13.4

L'HÔPITAL'S RULE

If the numerator and denominator both go to infinity (positive or negative), the limit remains the same after taking the derivative of the top and bottom. Using notation,

$$\lim_{x \to \pm\infty} \frac{f(x)}{g(x)} = \lim_{x \to \pm\infty} \frac{f'(x)}{g'(x)} \text{ if } \lim_{x \to \pm\infty} f(x) = \pm\infty \text{ and } \lim_{x \to \pm\infty} g(x) = \pm\infty$$

Example

Evaluate $\lim_{x \to \infty} \dfrac{\ln(x)}{1 - x}$.

Solution

Since $\lim_{x \to \infty} \ln(x) = \infty$ and $\lim_{x \to \infty}(1 - x) = -\infty$, we can use L'Hôpital's Rule.

$$\lim_{x \to \infty} \frac{\ln(x)}{1 - x} \overset{H}{=} \lim_{x \to \infty} \frac{\dfrac{d}{dx}(\ln(x))}{\dfrac{d}{dx}(1 - x)}$$

$$= \lim_{x \to \infty} \frac{\frac{1}{x}}{-1} = \lim_{x \to \infty} -\frac{1}{x} = 0$$

Note: The little H over the equals sign indicates that L'Hôpital's Rule has been used at that point of the computation. Examples like this demonstrate how $\ln(x)$ goes to infinity even slower than x does.

⚠ CAUTION

When applying L'Hôpital's Rule, we differentiate top and bottom separately and form the quotient of them. We do NOT apply the Quotient Rule.

Example

Evaluate $\lim_{x \to \infty} \dfrac{e^x}{x^3 + 2x^2 + 5x + 2}$.

Solution

Here, $\lim_{x \to \infty} e^x = \infty$ and $\lim_{x \to \infty}(x^3 + 2x^2 + 5x + 2) = \infty$, so we can use L'Hôpital's Rule.

$$\lim_{x \to \infty} \frac{e^x}{x^3 + 2x^2 + 5x + 2}$$

$$\overset{H}{=} \lim_{x \to \infty} \frac{\dfrac{d}{dx}(e^x)}{\dfrac{d}{dx}(x^3 + 2x^2 + 5x + 2)}$$

$$= \lim_{x \to \infty} \frac{e^x}{3x^2 + 4x + 5}$$

Here, we need to use L'Hôpital's Rule several more times:

$$\lim_{x \to \infty} \frac{e^x}{3x^2 + 4x + 5} \overset{H}{=} \lim_{x \to \infty} \frac{e^x}{6x + 4}$$

$$\overset{H}{=} \lim_{x \to \infty} \frac{e^x}{6} = \infty$$

This example shows how e^x is more powerful than x^3. If the denominator had an x^{100}, we'd have to use L'Hôpital's Rule 100 times, but in the end, e^x would drive everything to infinity.

Example

Evaluate $\lim\limits_{x\to-\infty} \dfrac{e^x}{x^5 + 7x - 1}$.

Solution

The limit $\lim\limits_{x\to-\infty} e^x = 0$ is not infinite. So we *can't* use L'Hôpital's Rule. The function e^x is only powerful when x goes to positive infinity. Instead, we use the old "plug in" method.

$$\lim_{x\to-\infty} \frac{e^x}{x^5 + 7x - 1} = \frac{0}{\text{something not zero}} = 0$$

Example

Evaluate $\lim\limits_{x\to\infty} \dfrac{\sin(x)}{x^2}$.

Solution

This has the same problem as the previous example. No matter what x may be, $\sin(x)$ will always be between -1 and 1. Thus, $-1 \le \sin(x) \le 1$ and so

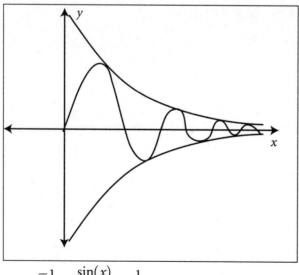

$$\frac{-1}{x^2} \le \frac{\sin(x)}{x^2} \le \frac{1}{x^2}$$

Because $\lim\limits_{x\to\infty} \dfrac{-1}{x^2} = 0$ and $\lim\limits_{x\to\infty} \dfrac{1}{x^2} = 0$, the function $\dfrac{\sin(x)}{x^2}$ is squeezed between them to zero as well: $\lim\limits_{x\to\infty} \dfrac{\sin(x)}{x^2} = 0$. This is called the *Squeeze Theorem* or the *Sandwich Theorem* because of the way $\dfrac{\sin(x)}{x^2}$ is squished between two curves, both going to zero.

Practice

Evaluate the following limits.

11. $\lim\limits_{x\to\infty} \dfrac{\ln(x^3)}{\ln(x) + 5}$

12. $\lim\limits_{x\to\infty} \dfrac{x + 5}{\sqrt{x} - 1}$

13. $\lim\limits_{x\to-\infty} \dfrac{x^2 + 5x - 10}{4x + 2}$

14. $\lim\limits_{x\to\infty} \dfrac{3x^2 + 2}{x - \ln(x)}$

15. $\lim\limits_{x\to\infty} \dfrac{e^{2x} + 3}{e^{3x} + 2}$

16. $\lim\limits_{x\to\infty} \dfrac{x + \ln(x)}{e^{2x}}$

17. $\lim\limits_{x\to\infty} \dfrac{x}{e^{-x}}$

18. $\lim\limits_{x\to\infty} \dfrac{\cos(x)}{x}$

19. $\lim\limits_{x\to\infty} \dfrac{4x^3 + 5x^2 + 2}{e^x - 7x^3}$

20. $\lim\limits_{x\to-\infty} \dfrac{4x^3 + 5x^2 + 2}{e^x - 7x^3}$

Sign Diagrams

In order to calculate the limits at vertical asymptotes, it is necessary to know where the function is positive and negative. The key is this: A continuous function cannot switch between positive and negative without being zero or undefined. Functions are zero when the top is zero *and* the bottom is NOT zero, and undefined where the denominator is zero. Mark all of these points on a number line. Between these points, the function must be entirely positive or negative. This can be found by testing any point in each interval.

For example, consider $f(x) = \dfrac{x - 4}{(x + 2)(1 - x)}$.

This function is zero at $x = 4$ and undefined at both $x = -2$ and $x = 1$. We mark these on a number line (see Figure 13.5).

In between $x = -2$ and $x = 1$, the function is either always positive or always negative. To find out which it is, we test a point between -2 and 1, such as 0. Because $f(0) = \dfrac{-4}{2(1)} = -2$ is negative, the function is always negative between -2 and 1. Similarly, we check a point between 1 and 4, such as $f(2) = \dfrac{-2}{4(-1)} = \dfrac{1}{2}$; a point after 4, such as $f(5) = \dfrac{1}{7(-4)} = -\dfrac{1}{28}$; and a point before -2, such as $f(-3) = \dfrac{-7}{-1(4)} = \dfrac{7}{4}$. The *sign diagram* for this function is shown in Figure 13.6.

This makes calculating the limits at the vertical asymptotes very easy. Not only does

$f(x) = \dfrac{x - 4}{(x + 2)(1 - x)}$ have vertical asymptotes at $x = -2$ and $x = 1$, but the limits are:

$$\lim_{x \to -2^-} \frac{x - 4}{(x + 2)(1 - x)} = \infty$$

$$\lim_{x \to -2^+} \frac{x - 4}{(x + 2)(1 - x)} = -\infty$$

$$\lim_{x \to 1^-} \frac{x - 4}{(x + 2)(1 - x)} = -\infty$$

$$\lim_{x \to 1^+} \frac{x - 4}{(x + 2)(1 - x)} = \infty$$

As before, we can calculate the limits at infinity:

$$\lim_{x \to \infty} \frac{x - 4}{(x + 2)(1 - x)} = \lim_{x \to \infty} \frac{x - 4}{-x^2 - x + 2} = 0$$

$$\lim_{x \to -\infty} \frac{x - 4}{(x + 2)(1 - x)} = 0$$

Note: We do not use a sign diagram when determining *horizontal* asymptotes.

Thus, $f(x)$ has a horizontal asymptote of $y = 0$. With all of this, we begin to get a picture of $f(x) = \dfrac{x - 4}{(x + 2)(1 - x)}$, which can be seen in Figure 13.7.

Figure 13.5

$f(x)$

Figure 13.6

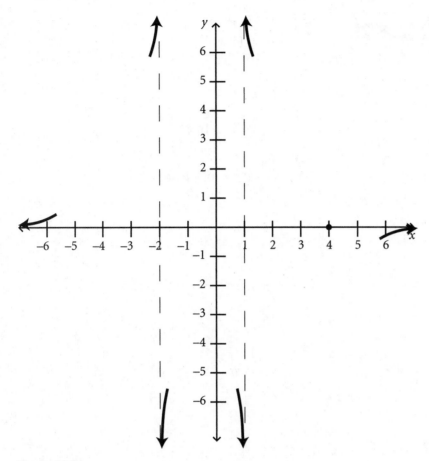

Figure 13.7

Notice that the horizontal asymptote $y = 0$ is approached from above as $x \to -\infty$, because $f(x)$ is always positive when $x < -2$. At the other end, the asymptote is approached from below as $x \to \infty$ because the function is negative when $x > 4$.

We shall deal with graphing more thoroughly in the next lesson.

Practice

For questions 21 through 26, determine all asymptotes, vertical and horizontal, of the following functions. Also, make a sign diagram for each.

21. $f(x) = \dfrac{x + 2}{x - 4}$

22. $g(x) = \dfrac{x - 3}{x^2 - 4}$

23. $h(x) = \dfrac{x^2 - 1}{(x + 3)^2}$

24. $k(x) = \dfrac{2x + 1}{x^2 - 4x + 3}$

25. $j(x) = \dfrac{-1}{(x^2 + 1)(x + 5)}$

26. $m(x) = \dfrac{x^6}{(x^2 + 9)(x + 1)(x + 2)^2}$

⚠ **CAUTION**

Remember, $x^2 + a^2 \neq (x - a)(x + a)$.

Evaluate the following limits.

27. $\lim\limits_{x \to 4^+} \dfrac{x + 2}{x - 4}$

28. $\lim\limits_{x \to 2^-} \dfrac{x - 3}{x^2 - 4}$

29. $\lim\limits_{x \to -3^-} \dfrac{x^2 - 1}{(x + 3)^2}$

30. $\lim\limits_{x \to 3^-} \dfrac{x + 1}{x^2 - 4x + 3}$

31. $\lim\limits_{x \to -5^+} \dfrac{-1}{(x^2 + 1)(x + 5)}$

32. $\lim\limits_{x \to -2} \dfrac{x^6}{(x^2 + 9)(x + 1)(x + 2)^2}$

14 ▶ USING CALCULUS TO GRAPH

Here is where everything comes together! We know how to find the domain, how to identify asymptotes, and how to plot points. With the help of the sign diagrams from the previous lesson, we shall be able to tell where a function is increasing and decreasing, and where it is concave up and down.

Quite simply, where the derivative is positive, the function is increasing. The derivative gives the slope of the tangent line at a point, and when this is positive, the function is heading upward, viewed from left to right. When the derivative is negative, the function slopes downward and decreases.

When the second derivative is positive, the function is concave up. This is because the second derivative says how the first derivative is changing. If the second derivative is positive, then the slopes are increasing. If the slopes, from left to right, increase from –2, to –1, to 0, to 1, to 2, and so on, then the graph must curve like the one in Figure 14.1. In other words, the curve must be concave up.

Similarly, if the second derivative is negative, the function curves downward like the one in Figure 14.2 and is concave down.

Figure 14.1

Figure 14.2

The concavity governs the shape of the graph, depending on whether the function $f(x)$ is increasing or decreasing. If $f(x)$ is increasing and concave up (thus, both $f'(x)$ and $f''(x)$ are positive), then the graph has the shape shown in Figure 14.3.

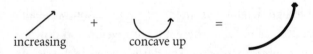

increasing + concave up =

Figure 14.3

If $f(x)$ is increasing and concave down (thus, $f'(x)$ is positive and $f''(x)$ is negative), then the graph has the shape shown in Figure 14.4.

increasing + concave down =

Figure 14.4

If $f(x)$ is decreasing and concave down (thus, both $f'(x)$ and $f''(x)$ are negative), then the graph has the shape shown in Figure 14.5.

decreasing + concave down =

Figure 14.5

If $f(x)$ is decreasing and concave up (thus, $f'(x)$ is negative and $f''(x)$ is positive), the graph has the shape of the one in Figure 14.6.

decreasing + concave up =

Figure 14.6

Example

Graph $f(x) = x^3 + 6x^2 - 15x + 10$.

Solution

This function is defined everywhere and thus has no vertical asymptotes. Because $\lim_{x \to \infty}(x^3 + 6x^2 - 15x + 10) = \infty$ and $\lim_{x \to -\infty}(x^3 + 6x^2 - 15x + 10) = -\infty$, there are no horizontal asymptotes.

The derivative $f'(x) = 3x^2 + 12x - 15 = 3(x^2 + 4x - 5) = 3(x + 5)(x - 1)$ is zero at $x = -5$ and $x = 1$. To form the sign diagram, we test: $f'(-6) = 21$, $f'(0) = -15$, and $f'(2) = 21$. **Note:** These points were chosen arbitrarily. Any point less than -5 will give the same information as the value $x = -6$, for instance, and any point between -5 and -1 will give the same information as the value at $x = 0$. Thus, the sign diagram for $f'(x)$ is shown in Figure 14.7.

Remember, the sign of $f'(x)$ determines whether $f(x)$ is increasing or decreasing.

Note: We use $f'(x)$ to see if the graph is increasing or decreasing, but $f(x)$ to find the y-value at a point.

Figure 14.7

Because the function increases up to $x = -5$ and then decreases immediately afterward, there is a local maximum at $x = -5$. The corresponding y-value is $y = f(-5) = 110$. Thus, $(-5,110)$ is a local maximum. Similarly, because the graph goes down to $x = 1$ and then goes up afterward, there is a local minimum at $x = 1$. The corresponding y-value is $f(1) = 2$, so $(1,2)$ is a local minimum.

A guideline for identifying local minimum and maximum points is shown in Figure 14.8.

The second derivative is $f''(x) = 6x + 12 = 6(x + 2)$, which is zero at $x = -2$. If we test the sign at $x = -3$ and $x = 0$, we get $f''(-3) = -6$ and $f''(0) = 12$. Thus, the sign diagram for $f''(x)$ is as shown in Figure 14.9.

Figure 14.8

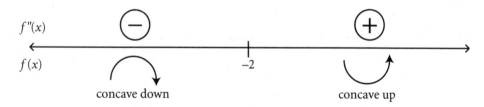

Figure 14.9

Clearly $x = -2$ is a point of inflection, because this is where the concavity switches from concave down to concave up. The y-value of this point is $f(-2) = 56$.

Before we draw the axes for the Cartesian plane, we should consider the three interesting points we have found: the local maximum at $(-5,110)$, the local minimum at $(1,2)$, and the point of inflection at $(-2,56)$. If our x-axis runs from $x = -10$ to $x = 10$, and our y-axis runs from 0 to 120, then all of these points can be plotted on our graph (see Figure 14.10).

Example

Graph $g(x) = \dfrac{x + 3}{x - 2}$.

Solution

The domain is $x \neq 2$. There is a vertical asymptote at $x = 2$. The sign diagram for $g(x)$ is shown in Figure 14.11.

Thus, $\displaystyle\lim_{x \to 2^-} \frac{x + 3}{x - 2} = -\infty$ and $\displaystyle\lim_{x \to 2^+} \frac{x + 3}{x - 2} = \infty$.

Because $\displaystyle\lim_{x \to \infty} \frac{x + 3}{x - 2} = 1$ and $\displaystyle\lim_{x \to -\infty} \frac{x + 3}{x - 2} = 1$,

$g(x)$ ⊕   ⊖   ⊕

above x-axis -3 below x-axis 2 above x-axis

Figure 14.11

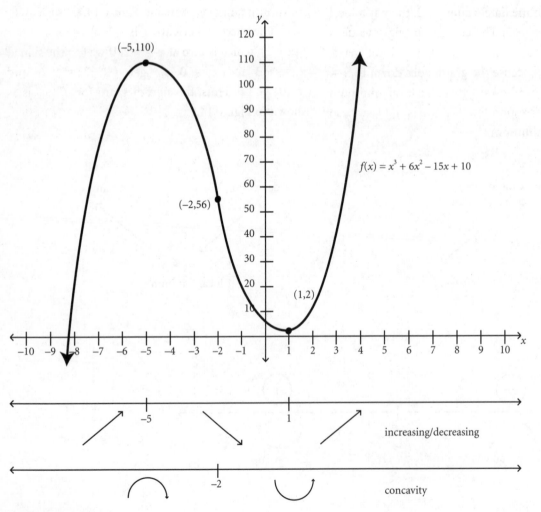

$f(x) = x^3 + 6x^2 - 15x + 10$

Figure 14.10

there is a horizontal asymptote at $y = 1$, both to the left and to the right. The derivative $g'(x) = \dfrac{1 \cdot (x - 2) - 1 \cdot (x + 3)}{(x - 2)^2} = \dfrac{-5}{(x - 2)^2}$ has the sign diagram shown in Figure 14.12.

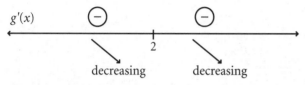

Figure 14.12

> ### ⚠ CAUTION
> Don't automatically assume that the signs in a sign diagram will alternate. In fact, they don't precisely when the number on the line comes from a factor raised to an *even* power, like $(x - 3)^2$.

The second derivative $g''(x) = \dfrac{10}{(x - 2)^3}$ has the sign diagram shown in Figure 14.13.

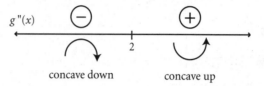

Figure 14.13

Because we have no points plotted at all, it makes sense to pick one or two to the left and right of the vertical asymptote at $x = 2$. At $x = 1$, $g(1) = -4$, so $(1, -4)$ is a point. At $x = 3$, $g(3) = 6$, so $(3,6)$ is another point. At $x = -3$, $g(-3) = 0$, so $(-3,0)$ is another nice point to know. Judging by these, it will be useful to have both the x- and y-axes run from -10 to 10.

To graph $g(x)$, it helps to start with the points and the asymptotes as shown in Figure 14.14.

Figure 14.14

Then we establish the shapes of the lines through these points using the concavity and the intervals of decrease (see Figure 14.15).

Example

Graph $h(x) = \dfrac{x^2 + 1}{x^2 - 1}$.

Solution

To start, $h(x) = \dfrac{x^2 + 1}{x^2 - 1} = \dfrac{x^2 + 1}{(x + 1)(x - 1)}$. Thus, $h(x)$ has vertical asymptotes at $x = 1$ and $x = -1$. The sign diagram for $h(x)$ is shown in Figure 14.16.

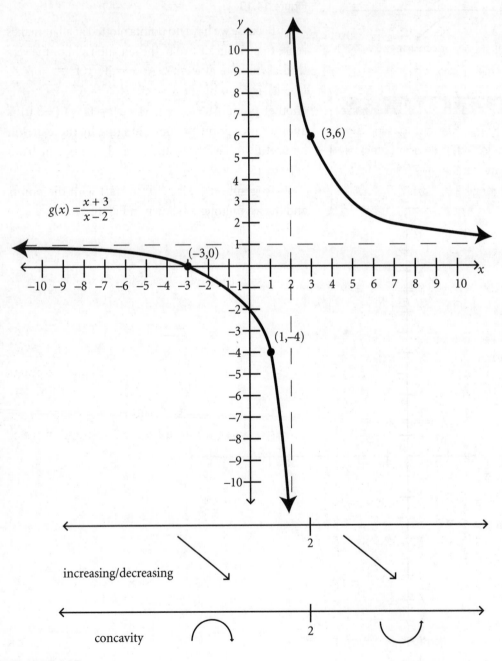

$g(x) = \dfrac{x + 3}{x - 2}$

increasing/decreasing

concavity

Figure 14.15

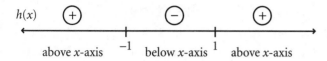

above x-axis −1 below x-axis 1 above x-axis

Figure 14.16

Note: $x^2 + 1$ can never be zero. The limits at the vertical asymptotes are thus:

$$\lim_{x \to -1^-} \frac{x^2 + 1}{x^2 - 1} = \infty$$

$$\lim_{x \to -1^+} \frac{x^2 + 1}{x^2 - 1} = -\infty$$

$$\lim_{x \to 1^-} \frac{x^2 + 1}{x^2 - 1} = -\infty$$

$$\lim_{x \to 1^+} \frac{x^2 + 1}{x^2 - 1} = \infty$$

Because $\lim\limits_{x \to \infty} \dfrac{x^2 + 1}{x^2 - 1} = 1$ and $\lim\limits_{x \to -\infty} \dfrac{x^2 + 1}{x^2 - 1} = 1$, there is a horizontal asymptote at $y = 1$.

The derivative is as follows:

$$h'(x) = \frac{2x(x^2 - 1) - 2x(x^2 + 1)}{(x^2 - 1)^2} = \frac{-4x}{(x - 1)^2(x + 1)^2}$$

Its sign diagram is shown in Figure 14.17. This indicates that there is a local maximum at $x = 0$. The corresponding y-value is $y = h(0) = -1$.

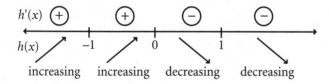

increasing   increasing   decreasing   decreasing

Figure 14.17

The second derivative is as follows:

$$h''(x) = \frac{-4(x^2 - 1)^2 - 2(x^2 - 1) \cdot 2x(-4x)}{(x^2 - 1)^4}$$

$$= \frac{-4(x^2 - 1) - 2 \cdot 2x(-4x)}{(x^2 - 1)^3}$$

$$= \frac{12x^2 + 4}{(x^2 - 1)^3} = \frac{12x^2 + 4}{(x - 1)^3(x + 1)^3}.$$

The sign diagram is shown in Figure 14.18. It looks like there ought to be points of inflection at $x = -1$ and $x = 1$, but these are asymptotes not in the domain, so there are no actual *points* where the concavity changes.

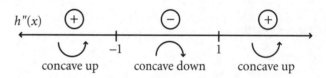

concave up   concave down   concave up

Figure 14.18

Before we graph the function, it will be useful to have a few more points. When $x = -2$, then $y = h(-2) = \frac{5}{3}$ and when $x = 2$, $y = h(2) = \frac{5}{3}$ as well. Thus, it will be useful to have the x- and y-axes run from about -3 to 3. We start with just the points and asymptotes (see Figure 14.19).

Then we add in the actual curves, guided by the concavity and the intervals of increase and decrease (see Figure 14.20).

Figure 14.19

Figure 14.20

Practice

Use the asymptotes, concavity, and intervals of increase and decrease, and concavity to graph the following functions.

1. $f(x) = x^2 - 30x + 10$

2. $g(x) = -4x - x^2$

3. $h(x) = 2x^3 - 3x^2 - 36x + 5$

4. $k(x) = 3x - x^3$

5. $f(x) = x^4 - 8x^3 + 5$

6. $g(x) = \dfrac{x}{x + 2}$

7. $h(x) = \dfrac{1}{x^2 - 9}$

8. $k(x) = \dfrac{x}{x^2 - 1}$

9. $j(x) = \dfrac{x^2 + 1}{x}$

10. $f(x) = \dfrac{x}{x^2 + 1}$

15 ▶ OPTIMIZATION

Knowing the minimum and maximum points of a function is useful for graphing and even more for solving real-life problems. Businesses want to maximize their profits, builders want to minimize their costs, drivers want to minimize distances, and people want to get the most for their money. If we can represent a situation with a function, then the derivative will help find optimal points.

If the derivative is zero or undefined at exactly one point, then this is very likely to be the optimal point. The *first derivative test* states that if the function increases before that point and decreases afterward, it is maximal (see Figure 15.1). Similarly, if the function decreases before the point and increases afterward, then the point is minimal.

The *second derivative test* states that if the second derivative is positive, then the function curves up, so a point of slope zero must be a minimum (see Figure 15.2). Similarly, if the second derivative is negative, the point of slope zero must be the highest point on the graph. Remember that we are assuming that only one point has slope zero or an undefined derivative.

If there are several points of slope zero and the function has a closed interval for a domain, then plug all the *critical points* (points of slope zero, points of undefined derivative, and the two endpoints of the interval) into the original function. The point with the highest y-value will be the absolute maximum, and the one with the smallest y-value will be the absolute minimum.

Figure 15.1

concave down

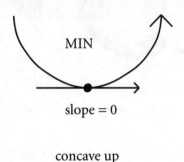

concave up

Figure 15.2

Example

A manager calculates that when x employees are working at the same time, the store makes a profit of $P(x) = 15x^2 - 48x - x^3$ dollars each hour. If there are ten employees and at least one must be working at any given time, how many employees should be scheduled to maximize profit?

Solution

This is an instance of a function defined on a closed interval because $1 \le x \le 10$ limits the options for x. The derivative of the profit function is $P'(x) = 30x - 48 - 3x^2 = -3(x^2 - 10x + 16) = -3(x - 2)(x - 8)$. Thus, the derivative is zero at $x = 2$ and at $x = 8$.

Because the function is defined on a closed interval, we cannot use the first or second derivative tests. Instead, we evaluate $f(x)$ at each of our critical points. These are the points of slope zero, $x = 2$ and $x = 8$, plus the endpoints of the interval, namely $x = 1$ and $x = 10$.

These are evaluated as follows: $P(1) = -34$, $P(2) = -44$, $P(8) = 64$, and $P(10) = 20$. If the manager wants to maximize the store profit, eight employees should be scheduled at the same time, because this will result in a maximal profit of \$64 each hour.

Example

A coffee shop owner calculates that if she sells cookies at \$$p$ each, she will sell $\dfrac{200}{p^2}$ cookies each day. If it costs her 20¢ to make each cookie, what price p will give her the greatest profit?

Solution

Profit is computed as: Profit = Revenue − Costs. If she charges \$$p$ per cookie, then she'll make and sell $\dfrac{200}{p^2}$ cookies each day. Thus, her revenue will be

$\left(\dfrac{200}{p^2}\right) \cdot p = \dfrac{200}{p}$ and her costs will be $\left(\dfrac{200}{p^2}\right) \cdot (0.20) = \dfrac{40}{p^2}$. Therefore, her profit function is

$P(p) = \dfrac{200}{p} - \dfrac{40}{p^2}$. We limit this to $p > 0.20$

because the only optimal situation would be when the cookies were sold for more than it cost to make them.

The derivative is $P'(p) = -\dfrac{200}{p^2} + \dfrac{80}{p^3}$, which is

zero when $\dfrac{80}{p^3} = \dfrac{200}{p^2}$ and therefore $80p^2 = 200p^3$, so

either $p = 0$ or $p = \dfrac{80}{200} = 0.40$. Because $p = 0$ is not

in the domain, the only place where the derivative is zero is at $p = 0.40$.

Using the first derivative test, we see that $P'(0.30) = 740$ and $P'(0.50) = -160$. So the sign diagram for P' is as shown in Figure 15.3. Thus, the absolute maximal profit occurs when the cookies are sold at 40¢ each.

Example

At \$1 per cup of coffee, a vendor sells 500 cups a day. When the price is increased to \$1.10, the vendor sells only 480 cups. If every 1¢ increase in price reduces the sales by two cups, what price per cup of coffee will maximize income?

Solution

Here, the income is Income = Price × Cups Sold. So if $x =$ the number of pennies by which the price is increased, then $I(x) = (1 + 0.01x) \cdot (500 - 2x)$. This simplifies to $I(x) = 500 + 3x - 0.02x^2$. And, the derivative is $I'(x) = 3 - 0.04x$. This is zero only when $x = \dfrac{3}{0.04} = 75$. The second derivative is $I''(x) = -0.04$, which is always negative, so $x = 75$ is maximal by the second derivative test. Thus, the maximal income will occur when the price is raised by $x = 75$¢ to \$1.75 per cup.

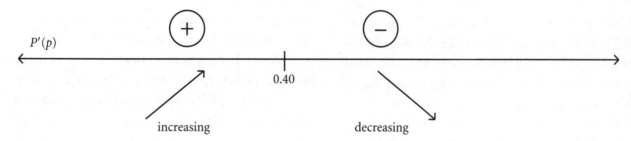

$P'(p)$

increasing   decreasing

Figure 15.3

Example

A farmer wants to build a rectangular pen with 80 feet of fencing. The pen will be built against the side of a barn, so one side won't need a fence. What dimensions will maximize the area of the pen? See Figure 15.4.

Figure 15.4

Solution

The area of the pen is Area $= x \cdot y$. We can't take the derivative yet because there are *two* variables. We need to use the additional information regarding how much fencing exists; there are 80 feet of fencing. Because no fencing will be required against the barn wall, the total lengths of the fence will be $y + x + y = 80$, thus $x = 80 - 2y$. We can plug this into the formula for area in order to obtain Area $= x \cdot y = (80 - 2y) \cdot y$. Now we have a function of one variable $A(y) = 80y - 2y^2$. The derivative is $A'(y) = 80 - 4y$. This is zero only when $y = 20$. Using the second derivative test, $A''(y) = -4$. So, the curve is concave down and the point $y = 20$ is the absolute maximum. The corresponding x-value is $x = 80 - 2y = 80 - 2(20) = 40$. Therefore, the pen with the maximal area will be $x = 40$ feet wide (along the barn) and $y = 20$ feet out from the barn wall.

Example

A manufacturer needs to design a crate with a square bottom and no top. It must hold exactly 32 cubic feet of shredded paper. What dimensions will minimize the material needed to make the crate (the surface area)? See Figure 15.5.

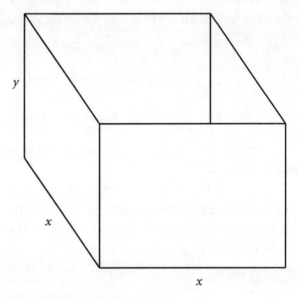

Figure 15.5

Solution

We want to minimize the surface area of the crate. The surface area of the box consists of four sides, each of area $x \cdot y$, plus the bottom, with an area of $x \cdot x = x^2$. Thus, the surface area is Area $= 4xy + x^2$. Again, we need to reduce this to a formula with only one variable in order to differentiate. We know that the volume must be 32 cubic feet, so Volume $= x^2 y = 32$. Thus, $y = \dfrac{32}{x^2}$. When we plug this into the surface area function, we get:

$$\text{Surface Area} = 4xy + x^2 = 4x\left(\frac{32}{x^2}\right) + x^2 = \frac{128}{x} + x^2.$$

So, we have a function of one variable $A(x) = \dfrac{128}{x} + x^2$.

The derivative is:

$$A'(x) = -\frac{128}{x^2} + 2x.$$

which is zero when

$$-\frac{128}{x^2} + 2x = 0 \text{ or } x^3 = 64, \text{ so } x = 4.$$

The second derivative is:

$$A''(x) = \frac{256}{x^3} + 2,$$

which is positive when $x = 4$. So, the curve is concave up and the sole point of slope zero is the absolute minimum. Thus, the surface area of the crate will be minimized if $x = 4$ feet and $y = \frac{32}{x^2} = \frac{32}{4^2} = 2$ feet.

Practice

1. Suppose a company makes a profit of $P(x) = \frac{1,000}{x} - \frac{5,000}{x^2} + 100$ dollars when it makes and sells $x > 0$ items. How many items should it make to maximize profit?

2. When 30 orange trees are planted on an acre, each will produce 500 oranges a year. For every additional orange tree planted, each tree will produce 10 fewer oranges. How many trees should be planted to maximize the yield?

3. An artist can sell 20 copies of a painting at $100 each, but for each additional copy she makes, the value of each painting will go down by a dollar. Thus, if 22 copies are made, each will sell for $98. How many copies should she make to maximize her sales?

4. A garden has 200 pounds of watermelons growing in it. Every day, the total amount of watermelon increases by 5 pounds. At the same time, the price per pound of watermelon goes down by 1¢. If the current price is 90¢ per pound, how much longer should the watermelons grow in order to fetch the highest price possible?

5. A farmer has 400 feet of fencing to make three rectangular pens. What dimensions x and y will maximize the total area?

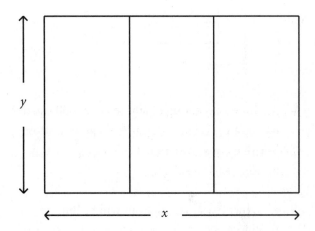

6. Four rectangular pens will be built along a river by using 150 feet of fencing. What dimensions will maximize the area of the pens?

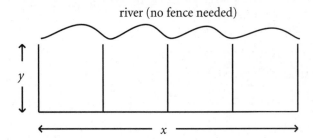
river (no fence needed)

7. The surface area of a can is Area = $2\pi r^2 + 2\pi rh$, where the height is h and the radius is r. The volume is Volume = $\pi r^2 h$. What dimensions minimize the surface area of a can with volume 16π cubic inches?

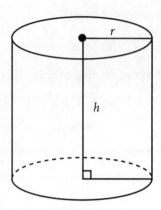

8. A painter has enough paint to cover 600 square feet of area. What is the largest square-bottom box that could be painted (including the top, bottom, and all sides)?

9. A box with a square bottom will be built to contain 40,000 cubic feet of grain. The sides of the box cost 10¢ per square foot to build, the roof costs $1 per square foot to build, and the bottom will cost $7 per square foot to build. What dimensions will minimize the building costs?

10. A printed page will have a total area of 96 square inches. The top and bottom margins will be 1 inch each, and the left and right margins will be $\frac{3}{2}$ inches each. What overall dimensions for the page will maximize the area of the space inside the margins?

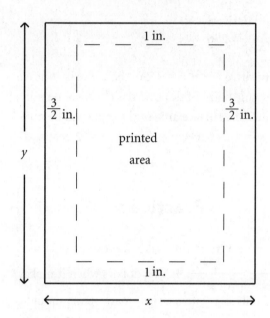

16 ▶ THE INTEGRAL AND AREAS UNDER CURVES

Around the same time that many great mathematicians focused on figuring out the slopes of tangent lines, other mathematicians were working on an entirely different problem. They wanted to be able to compute the area underneath *any* curve $y = f(x)$, such as the one shown in Figure 16.1.

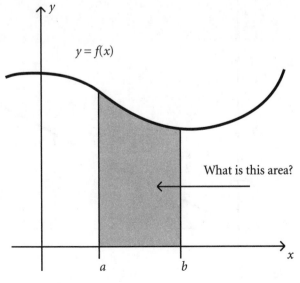

$y = f(x)$

What is this area?

Figure 16.1

The curvy nature of the upper curve $y = f(x)$ presents a problem when finding area. But we know how to find areas of rectangles. Therefore, the approach we shall take is to *approximate* the region using better and better rectangular staircases, as follows.

Number of Rectangles	Diagram	Area of rectangular staircase (Dashed Portion of the Region)
1		Area = $\underbrace{f(5)}_{\text{height}} \cdot \underbrace{4}_{\text{base}}$
		Bad approximation of the area of the original region.
2		Area = $\underbrace{f(3)}_{\text{height}} \cdot \underbrace{2}_{\text{base}} + \underbrace{f(5)}_{\text{height}} \cdot \underbrace{2}_{\text{base}}$
		Better approximation than with 1 rectangle, but not great.
4		Area = $\underbrace{f(2)}_{\text{height}} \cdot \underbrace{1}_{\text{base}} + \underbrace{f(3)}_{\text{height}} \cdot \underbrace{1}_{\text{base}}$ $+ \underbrace{f(4)}_{\text{height}} \cdot \underbrace{1}_{\text{base}} + \underbrace{f(5)}_{\text{height}} \cdot \underbrace{1}_{\text{base}}$
		Even better approximation.

Figure 16.2

If we keep using more rectangles to dissect the original region, the rectangles gobble up more of the region, and so the approximation to the actual area is better. The number to which these approximate areas get close as we do so is called the integral of f on $[a,b]$, and is denoted $\int_a^b f(x)\,dx$.

Note: The "elongated S" symbol (called the *integral sign*), $\int$, is used because the process involves a sum. The integrand $f(x)\,dx$ resembles the form of what we are adding, namely height $\times$ base. The dx portion doesn't actually enter into the computation, though.

Example

Evaluate the integral $\int_0^4 \frac{1}{2}x\,dx$.

Solution

This represents the area between the curve $y = \frac{1}{2}x$, the x-axis, the line $x = 0$, and the line $x = 4$ (see Fig-

ure 16.3). This region happens to be a triangle with a height of 2 and a base of 4. The area of the triangle is $\frac{1}{2}(2)(4) = 4$. Thus, $\int_0^4 \frac{1}{2}x\,dx = 4$.

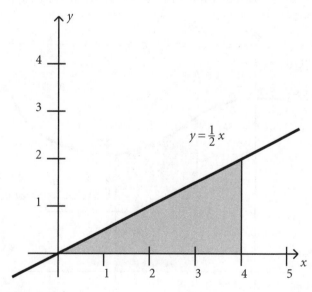

Figure 16.3

If the region is *below* the x-axis, the area is counted negatively. Therefore, really $\int_a^b f(x)\,dx$ represents "the area between the curve $y = f(x)$, the x-axis, $x = a$, and $x = b$, where area below the x-axis is counted negatively."

Example

Evaluate the integrals $\int_1^3 f(x)\,dx$, $\int_1^4 f(x)\,dx$, $\int_1^6 f(x)\,dx$, and $\int_1^3 f(x)\,dx$ where the graph of $y = f(x)$ is shown in Figure 16.4.

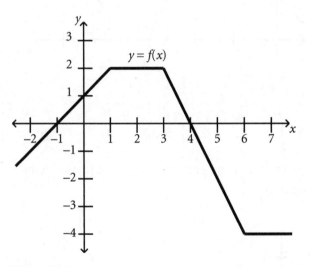

Figure 16.4

Solution

First, $\int_1^3 f(x)\,dx = 2(2) = 4$ because this area is a square above the x-axis (see Figure 16.5).

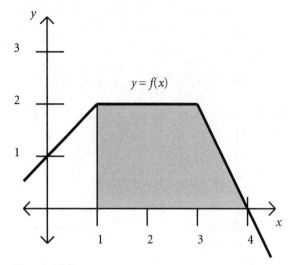

Figure 16.5

Next, $\int_1^4 f(x)\,d = 2(2) + \frac{1}{2} \cdot (1) \cdot 2 = 5$ because this area is a square plus a triangle (see Figure 16.6).

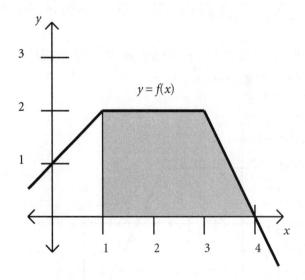

Figure 16.6

For $\int_1^6 f(x)\,dx$, we must calculate how much area is above the x-axis and how much is below (see Figure 16.7).

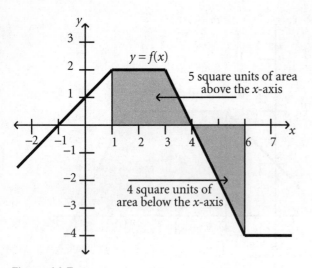

Figure 16.7

There are 5 units of area above the x-axis and 4 units below, so $\int_{1}^{6} f(x)\,dx = 5 - 4 = 1$.

Finally, $\int_{6}^{7} f(x)\,dx$ represents a rectangle of area 4 that is entirely below the x-axis. Thus, $\int_{1}^{6} f(x)\,dx = (-4)(1) = -4$ (see Figure 16.8).

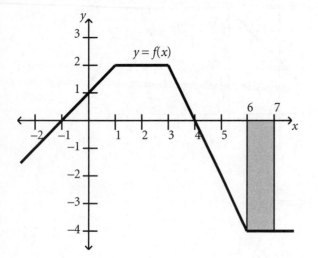

Figure 16.8

Practice

Evaluate the following integrals.

Use the following graph to solve practice problems 1, 2, and 3.

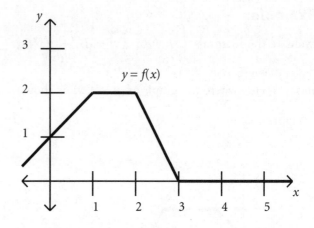

1. $\int_{1}^{2} f(x)\,dx$ **2.** $\int_{0}^{1} f(x)\,dx$ **3.** $\int_{0}^{2} f(x)\,dx$

Use the following graph to solve practice problems 4, 5, and 6.

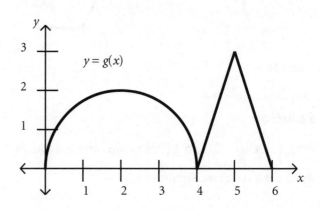

4. $\int_{0}^{4} g(x)\,dx$ **5.** $\int_{4}^{6} g(x)\,dx$ **6.** $\int_{0}^{6} g(x)\,dx$

Use the following graph to solve practice problems 7, 8, and 9.

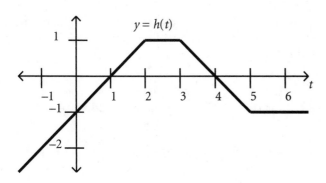

7. $\displaystyle\int_{-1}^{6} h(t)\,dt$ **8.** $\displaystyle\int_{-1}^{4} h(t)\,dt$ **9.** $\displaystyle\int_{4}^{6} h(t)\,dt$

Use the following graph to solve practice problems 10 through 12.

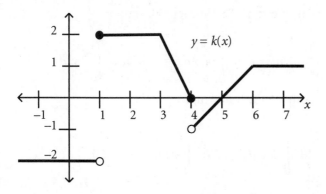

10. $\displaystyle\int_{0}^{7} k(x)\,dx$ **11.** $\displaystyle\int_{4}^{6} k(x)\,dx$ **12.** $\displaystyle\int_{4}^{5} k(x)\,dx$

For questions 13 through 18, compute the integral.

13. $\displaystyle\int_{0}^{4} (x+2)\,dx$ **14.** $\displaystyle\int_{1}^{4} 2\,dx$ **15.** $\displaystyle\int_{1}^{5} (t-3)\,dt$

16. $\displaystyle\int_{-2}^{5} x\,dx$ **17.** $\displaystyle\int_{1}^{6} 2x\,dx$ **18.** $\displaystyle\int_{0}^{8} (2x-2)\,dx$

You might have noticed that:

$$\int_a^b f(x)\,dx + \int_b^c f(x)\,dx = \int_a^c f(x)\,dx$$

The area between a and c is the area from a to b plus the area from b to c, assuming, of course, that $a < b < c$ (see Figure 16.9).

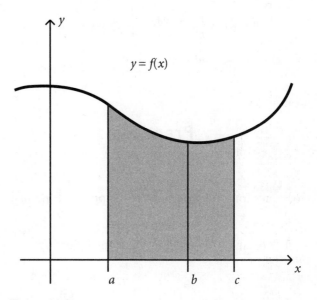

Figure 16.9

Similarly, $\displaystyle\int_a^c f(x)\,dx - \int_b^c f(x)\,dx = \int_a^b f(x)\,dx.$

We can use these to perform calculations, even when the exact functions are unknown.

Example

If $\displaystyle\int_{3}^{5} f(x)\,dx = 7$ and $\displaystyle\int_{3}^{10} f(x)\,dx = 15$, then what is

$\displaystyle\int_{3}^{10} f(x)\,dx$?

Solution

$$\int_{3}^{10} f(x)\,dx = \int_{3}^{5} f(x)\,dx + \int_{3}^{10} f(x)\,dx$$
$$= 7 + 15 = 22$$

Example

If $\displaystyle\int_0^{10} g(x)\,dx = 38$ and $\displaystyle\int_8^{10} g(x)\,dx = -12$, then what is $\displaystyle\int_0^8 g(x)\,dx$?

Solution

$$\int_0^8 g(x)\,dx = \int_0^{10} g(x)\,dx - \int_8^{10} g(x)\,dx$$
$$= 38 - (-12) = 50$$

Practice

For questions 19 through 21, suppose $\displaystyle\int_0^6 f(x)\,dx,$ $\displaystyle\int_6^7 f(x)\,dx,$ and $\displaystyle\int_6^{11} f(x)\,dx.$ Evaluate the following.

19. $\displaystyle\int_0^7 f(x)\,dx$ **20.** $\displaystyle\int_6^{11} f(x)\,dx$ **21.** $\displaystyle\int_0^{11} f(x)\,dx$

For questions 22 through 24, suppose $\displaystyle\int_1^{14} g(t)\,dt = -3,$ $\displaystyle\int_{10}^{14} g(t)\,dt = 8,$ and $\displaystyle\int_1^5 g(t)\,dt = -10.$ Evaluate the following.

22. $\displaystyle\int_5^{14} g(t)\,dt$ **23.** $\displaystyle\int_1^{10} g(t)\,dt$ **24.** $\displaystyle\int_5^{10} g(t)\,dt$

For questions 25 through 27, suppose $\displaystyle\int_{-2}^{11} h(x)\,dx = 20,$ $\displaystyle\int_{-2}^1 h(x)\,dx = 12,$ and $\displaystyle\int_{-2}^{10} h(x)\,dx = -5.$ Evaluate the following.

25. $\displaystyle\int_1^{11} h(x)\,dx$ **26.** $\displaystyle\int_1^{10} h(x)\,dx$ **27.** $\displaystyle\int_{10}^{11} h(x)\,dx$

For questions 28 through 30, suppose $\displaystyle\int_{-1}^0 j(x)\,dx = 2,$ $\displaystyle\int_0^1 j(x)\,dx = 3,$ $\displaystyle\int_1^2 j(x)\,dx = -4,$ and $\displaystyle\int_{-1}^3 j(x)\,dx = 12.$ Evaluate the following.

28. $\displaystyle\int_2^3 j(x)\,dx$ **29.** $\displaystyle\int_0^2 j(x)\,dx$ **30.** $\displaystyle\int_{-1}^2 j(x)\,dx$

17 ▶ THE FUNDAMENTAL THEOREM OF CALCULUS

Here comes the resounding climax of calculus. It would be best to read this lesson with some bombastic orchestral music like that of Wagner or Orff! The initial question here is innocent enough: If we make a function using that "area under a curve" stuff, what would its derivative be? To make this precise, suppose that our curve is $y = f(t)$ (see Figure 17.1). We use the variable t in order to save x for something else later.

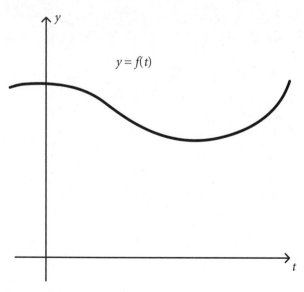

Figure 17.1

Now let our "area under the curve function" be $g(x)$ = the area under the curve $y = f(t)$ between 0 and some point x. That is, $g(x) = \int_0^x f(t)\,dt$. This area is illustrated in Figure 17.2.

Figure 17.3

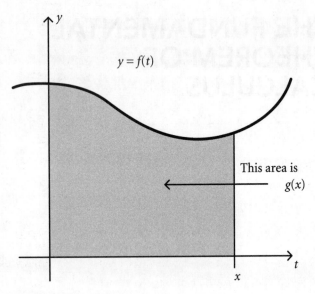

Figure 17.2

Example

If $f(t) = 2t$ and $g(x) = \int_0^x f(t)\,dt$, then what is $g(3)$?

Solution

$g(3) = \int_0^3 f(t)\,dt = \int_0^3 2t\,dt$ = the area beneath the curve $y = 2t$ from 0 to 3. The graph of $f(t) = 2t$ and the region are shown in Figure 17.3. This region is a triangle with base 3 and height 6, so

$g(3) = \int_0^3 2t\,dt = \frac{1}{2}(3)(6) = 9$.

Practice

For questions 1 through 6, suppose $f(t) = 3t + 1$ and $g(x) = \int_0^x f(t)\,dt$. Evaluate the following.

(**Hint:** The area of a trapezoid with bases b_1 and b_2 and height h is $A = \frac{1}{2}h(b_1 + b_2)$.

1. $g(0)$

2. $g(1)$

3. $g(2)$

4. $g(3)$

5. $g(4)$

6. $g(5)$

For questions 7 through 12, suppose $f(t) = 7$ and $g(x) = \int_0^x f(t)\,dt$. Evaluate the following.

7. $g(0)$

8. $g(1)$

9. $g(2)$

10. $g(3)$

11. $g(4)$

12. $g(5)$

Now that you are familiar with $g(x) = \int_0^x f(t)\,dt$, we can answer the next question: What is the derivative of $g(x)$?

Begin with the definition of the derivative.

$$g'(x) = \lim_{h \to 0} \frac{g(x+h) - g(x)}{h}$$

Use $g(x) = \int_0^x f(t)\,dt$.

$$g'(x) = \lim_{h \to 0} \frac{\int_0^{x+h} f(t)\,dt - \int_0^x f(t)\,dt}{h}$$

Use $\int_a^c f(t)\,dt - \int_a^b f(t)\,dt = \int_b^c f(t)\,dt$.

$$g'(x) = \lim_{h \to 0} \frac{\int_x^{x+h} f(t)\,dt}{h}$$

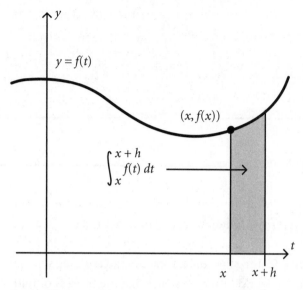

Figure 17.4

Now the integral $\int_x^{x+h} f(t)\,dt$ represents the skinny little area just to the right of point x (see Figure 17.4). This is *almost* a rectangle with a base of h and a height of $f(x)$, so the integral $\int_x^{x+h} f(t)\,dt$ is *almost* $h \cdot f(x)$. As h goes to zero, this approximation gets better. Therefore,

$$g'(x) = \lim_{h \to 0} \frac{\int_x^{x+h} f(t)\,dt}{h}$$

$$= \lim_{h \to 0} \frac{h \cdot f(x)}{h} = \lim_{h \to 0} f(x) = f(x)$$

THE FUNDAMENTAL THEOREM OF CALCULUS

The Fundamental Theorem of Calculus can be written as follows:

$$\int_a^b f(x)\,dx = \int_0^b f(x)\,dx - \int_0^a f(x)\,dx = g(b) - g(a), \text{ where } g'(x) = f(x)$$

What does this mean? It means that the derivative of the function $g(x)$, which represents "the area under the curve from 0 to x," is the very function $f(x)$ used to draw the curve. It came as an amazing surprise to the world of mathematics that the process of finding the slope of a tangent line and the process of finding the area under a curve were such inverses. In order to find the area under a curve $y = f(x)$, we need to find a function $g(x)$ whose derivative is $f(x)$.

We can use this to evaluate $\int_a^b f(x)\,dx$ using the Fundamental Theorem of Calculus.

For example, the derivative of $g(x) = x^2$ is $g'(x) = 2x$. Thus, the area under $f(x) = 2x$ between $x = 3$ and $x = 5$ is $\int_3^5 2x\,dx = g(5) - g(3) = 5^2 - 3^2 = 16$. This is exactly the area of the trapezoid under the line $y = 2x$ between $x = 3$ and $x = 5$.

Example

The derivative of $g(x) = \frac{1}{3}x^3$ is $g'(x) = x^2$. Use this to evaluate $\int_{-1}^2 x^2\,dx$.

Solution

By the Fundamental Theorem of Calculus,

$$\int_a^b f(x)\,dx = g(b) - g(a), \text{ where } g'(x) = f(x). \text{ Thus,}$$

$$\int_{-1}^2 x^2\,dx = g(2) - g(-1)$$

$$= \frac{1}{3}(2)^3 - \frac{1}{3}(-1)^3 = \frac{8}{3} + \frac{1}{3} = 3.$$

Even if we *had* drawn out the graph of $y = x^2$, how would we have been able to guess that the area of the two shaded curves add up to exactly three? This is why the Fundamental Theorem of Calculus is so powerful! (See Figure 17.5.)

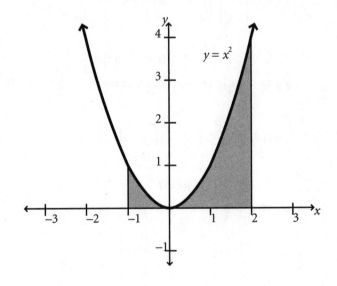

Figure 17.5

Example

If $g(x) = x^4$, then $g'(x) = 4x^3$. Use this to evaluate $\int_{-1}^{1} 4x^3 \, dx$.

Solution

$$\int_{-1}^{1} 4x^3 \, dx = g(1) - g(-1)$$

$$= 1^4 - (-1)^4 = 1 - 1 = 0$$

The answer is zero because there is exactly as much area above the x-axis (which counts positively) as there is below the x-axis (which counts negatively).

Practice

For questions 13 through 16, use $\dfrac{d}{dx}(x^2 + x) = 2x + 1$ to evaluate the following.

13. $\int_{1}^{3} (2x + 1) \, dx$

14. $\int_{-3}^{1} (2x + 1) \, dx$

15. $\int_{2}^{6} (2x + 1) \, dx$

16. $\int_{0}^{4} (2x + 1) \, dx$

For questions 17 through 20, use $\dfrac{d}{dx}\left(\dfrac{2}{3}x^{\frac{3}{2}}\right) = \sqrt{x}$ to evaluate the following.

17. $\int_{0}^{1} \sqrt{x} \, dx$

18. $\int_{0}^{4} \sqrt{x} \, dx$

19. $\int_{4}^{9} \sqrt{x} \, dx$

20. $\int_{0}^{100} \sqrt{x} \, dx$

For questions 21 through 24, use $\dfrac{d}{dx}(\sin(x)) = \cos(x)$ to evaluate the following.

21. $\int_{0}^{\frac{\pi}{4}} \cos(x) \, dx$

22. $\int_{\frac{\pi}{6}}^{\frac{\pi}{2}} \cos(x) \, dx$

23. $\int_{\frac{3\pi}{4}}^{\pi} \cos(x) \, dx$

24. $\int_{\frac{3\pi}{2}}^{2\pi} \cos(x) \, dx$

18 ▶ ANTIDIFFERENTIATION

The Fundamental Theorem of Calculus shows that the area under the curve, $\displaystyle\int_a^b f(x)\,dx$, can be calculated using a function $g(x)$ whose derivative is $g'(x) = f(x)$. Symbolically,

$$\int_a^b f(x)\,dx = \left[g(x)\right]_a^b = g(b) - g(a)$$

Because of this, the symbol $\displaystyle\int$, without the limits of integration, is used to represent the opposite of taking the derivative. An integral like $\displaystyle\int_a^b f(x)\,dx$ is called a *definite integral* because it represents a definite area.

An integral like $\displaystyle\int f(x)\,dx$ is called an *indefinite integral* because it represents another function.

Thus, $\displaystyle\int f(x)\,dx$ means the *antiderivative* of $f(x)$ or "the function whose derivative is $f(x)$." For example, $\displaystyle\int 2x\,dx$ asks "whose derivative is $2x$?" This could be x^2 because $\dfrac{d}{dx}(x^2) = 2x$. However, it could also be $x^2 + 5$ because $\dfrac{d}{dx}(x^2 + 5) = 2x$. In fact, because the derivative of a constant is zero, $\displaystyle\int 2x\,dx$ could be x^2 plus any constant. Therefore, we write $\displaystyle\int 2x\,dx = x^2 + c$ where c is any constant.

The brackets $\left[\ldots\right]_a^b$ are just a way of keeping track of the *limits of integration a* and *b* before they are plugged into $g(x)$ and subtracted.

We usually simply write $\int 2x\,dx = x^2 + c$ without actually saying that c stands for "some constant." In many ways, the "plus c" is the trademark of the indefinite integral because every problem that begins with $\int (\ldots.)\,dx$ ends with $+\,c$.

If we are dealing with a definite integral like $\int_3^5 2x\,dx$, then it does not matter what constant we use. For example:

$$\int_3^5 2x\,dx = \left[x^2 + c\right]_3^5$$

$$= (5^2 + c) - (3^2 + c)$$

$$= 25 + \cancel{c} - 9 - \cancel{c} = 16$$

The "plus c" will always cancel out in the subtraction. As such, we simply use $c = 0$ and write:

$$\int_3^5 2x\,dx = \left[x^2\right]_3^5 = 5^2 - 3^2 = 25 - 9 = 16$$

Example

Use $\dfrac{d}{dx}(x^3 + 10x^2 + 3x) = 3x^2 + 20x + 3$ to evaluate $\int (3x^2 + 20x + 3)\,dx$ and $\int_1^2 (3x^2 + 20x + 3)\,dx$.

Solution

Because $\dfrac{d}{dx}(x^3 + 10x^2 + 3x) = 3x^2 + 20x + 3$, we know that:

$$\int (3x^2 + 20x + 3)\,dx = x^3 + 10x^2 + 3x + c$$

Also,

$$\int_1^2 (3x^2 + 20x + 3)\,dx = \left[x^3 + 10x^2 + 3x\right]_1^2$$

$$= ((2)^3 + 10\cdot(2)^2 + 3\cdot(2)) -$$

$$((1)^3 + 10\cdot(1) + 3\cdot(1))$$

$$= 8 + 40 + 6 - (1 + 10 + 3) = 54 - 14 = 40$$

The general process for finding antiderivatives of powers is fairly simple. To take the derivative of $f(x) = x^5$, we first multiply by the exponent 5, and then we subtract one from the exponent. Thus, $f'(x) = 5x^4$.

To *antidifferentiate* $\int 5x^4 dx$, we must do the exact opposite of this process. First, we add one to the exponent, and then we divide the result by the new exponent. Thus, $\int 5x^4 dx = \dfrac{5x^{4+1}}{4+1} + c = x^5 + c$. In general, we write:

$$\int x^n dx = \frac{x^{n+1}}{n+1} + c \text{ if } n \neq -1$$

You can verify your answer by taking its derivative. If the derivative of your answer is what you were trying to integrate, then you are correct.

The derivative of $\frac{2}{3}x^{\frac{3}{2}} + c$ is $\frac{d}{dx}\left(\frac{2}{3}x^{\frac{3}{2}} + c\right) = \frac{2}{3} \cdot \frac{3}{2}x^{\frac{1}{2}} + 0 = x^{\frac{1}{2}} = \sqrt{x}$. This verifies that

$$\int \sqrt{x}\,dx = \frac{2}{3}x^{\frac{3}{2}} + c.$$

Example

Evaluate $\int x^7 dx$.

Solution

$$\int x^7 dx = \frac{x^{7+1}}{7+1} + c = \frac{1}{8}x^8 + c$$

Example

Evaluate $\int \sqrt{x}\,dx$

Solution

$$\int \sqrt{x}\,dx = \int x^{\frac{1}{2}}dx$$

$$= \frac{x^{\frac{1}{2}+1}}{\frac{1}{2}+1} + c = \frac{x^{\frac{3}{2}}}{\frac{3}{2}} + c = \frac{2}{3}x^{\frac{3}{2}} + c$$

Example

Evaluate $\int_0^2 x^3 dx$.

Solution

$$\int_0^2 x^3 dx = \left[\frac{1}{4}x^4\right]_0^2$$

$$= \frac{1}{4} \cdot 2^4 - \frac{1}{4} \cdot 0^4$$

$$= \frac{1}{4} \cdot 16 - \frac{1}{4} \cdot 0 = 4 - 0 = 4$$

Practice

Evaluate the following integrals.

1. $\int x^4 dx$

2. $\int x^{12} dx$

3. $\int u^6 du$

4. $\int_{-1}^{1} x^5 dx$

5. $\int_{-1}^{9} x\,dx$

6. $\int t^{-3} dt$

$$\int_a^b cf(x)dx = c\int_a^b f(x)dx, \int_a^b (f(x) + g(x))dx = \int_a^b f(x)dx + \int_a^b g(x)dx$$

7. $\displaystyle\int_{-4}^{-1} t^{-2}dt$

8. $\displaystyle\int x^{\frac{5}{3}}dx$

9. $\displaystyle\int \sqrt[4]{x}\, dx$

10. $\displaystyle\int \sqrt[3]{u}\,du$.

11. $\displaystyle\int 5dx$

12. $\displaystyle\int_{-1}^{4} 8dx$

Just as with derivatives, constants can stand aside, and the terms of sums can be dealt with separately.

Example

Evaluate $\displaystyle\int 5x^2dx$.

Solution

$$\int 5x^2dx = 5\int x^2dx = 5\left(\frac{1}{3}x^3 + c\right) = \frac{5}{3}x^3 + c$$

Example

Evaluate $\displaystyle\int (2t^3 - 8t + 7)dt$.

Solution

$$\int (2t^3 - 8t + 7)dt = 2\int t^3dt - 8\int tdt + \int 7dt$$

$$= 2\cdot\frac{1}{4}t^4 - 8\cdot\frac{1}{2}t^2 + 7t + c$$

$$= \frac{1}{2}t^4 - 4t^2 + 7t + c$$

Example

Evaluate $\displaystyle\int_1^4 \left(6\sqrt{x} - \frac{8}{x^5}\right) dx$.

Solution

It always helps to first write everything in exponential form.

$$\int_1^4 \left(6\sqrt{x} - \frac{8}{x^5}\right) dx = \int_1^4 (6x^{\frac{1}{2}} - 8x^{-5})\, dx$$

$$= \left[6\cdot\left(\frac{2}{3}x^{\frac{3}{2}}\right) - 8\left(\frac{x^{-4}}{-4}\right)\right]_1^4$$

$$= \left[4x^{\frac{3}{2}} + \frac{2}{x^4}\right]_1^4$$

$$= \left(4(8) + \frac{2}{256}\right) - \left(4 + \frac{2}{1}\right)$$

$$= 26 + \frac{1}{128} = \frac{3,329}{128}$$

Practice

Evaluate the following integrals.

13. $\displaystyle\int 9x^4 dx$

14. $\displaystyle\int 8u^2 du$

15. $\displaystyle\int \left(\sqrt[3]{x} - \sqrt{x}\right) dx$

16. $\displaystyle\int (6x^2 - 10x + 5) dx$

17. $\displaystyle\int_0^2 12x^3 dx$

18. $\displaystyle\int_1^2 (3x^2 + 4)\, dx$

19. $\displaystyle\int (3t^{11} + 9t^2 + t)\, dt$

20. $\displaystyle\int_0^3 (1 - t^2)\, dt$

21. $\displaystyle\int \left(2t^{-3} - 4t^{-2} - t\right) dt$

22. $\displaystyle\int_0^2 (10u^4 - 4u + 1)\, du$

23. $\displaystyle\int_1^9 12\sqrt{x}\, dx$

24. $\displaystyle\int \left(3x^{\frac{10}{3}} - 8x^{\frac{4}{7}}\right) dx$

The antiderivatives of e^x, $\sin(x)$, and $\cos(x)$ follow directly from their derivatives:

$$\int e^x dx = e^x + c \text{ because } \frac{d}{dx}(e^x) = e^x$$

$$\int \cos(x) dx = \sin(x) + c \text{ because}$$

$$\frac{d}{dx}(\sin(x)) = \cos(x)$$

$$\int_0^2 (10u^4 - 4u + 1)\, du \text{ because}$$

$$\frac{d}{dx}(-\cos(x)) = \sin(x)$$

The integral of $\ln(x)$ will have to wait until Lesson 20, though we can use the fact that $\frac{d}{dx}(\ln(x)) = \frac{1}{x}$ right now. We are inclined to say that $\int \frac{1}{x} dx = \ln(x) + c$, but this is not entirely correct. The derivative of $\ln(-x)$ is $\frac{d}{dx}(\ln(-x)) = \frac{1}{-x} \cdot \frac{d}{dx}(-x)$ $= \frac{1}{-x} \cdot (-1) = \frac{1}{x}$ as well. It does not matter if the x inside the natural logarithm is positive or negative, so we can generalize with the absolute value $|x|$.

$$\int \frac{1}{x} dx = \ln|x| + c$$

Incidentally, this nicely fills a hole in an earlier formula:

$$\int x^n dx = \frac{x^{n+1}}{n+1} + c \text{ if } n \neq -1$$

and if $n = -1$ then

$$\int x^{-1} dx = \int \frac{1}{x} dx = \ln |x| + c$$

Example

Evaluate $\int (3\sin(x) + 5\cos(x)) dx$.

Solution

$$\int (3\sin(x) + 5\cos(x)) dx =$$

$$-3\cos(x) + 5\sin(x) + c$$

Example

Evaluate $\int_0^1 (3t^2 - 5e^t) \, dt$.

Solution

$$\int_0^1 (3t^2 - 5e^t) \, dt = [t^3 - 5e^t]_0^1$$

$$= (1^3 - 5e^1) - (0^3 - 5e^0)$$

$$= 1 - 5e + 5 = 6 - 5e$$

Example

Evaluate $\int \left(x^2 + x + 1 + \frac{1}{x} + \frac{1}{x^2} \right) dx$.

Solution

$$\int \left(x^2 + x + 1 + \frac{1}{x} + \frac{1}{x^2} \right) dx =$$

$$\int \left(x^2 + x^1 + x^0 + x^{-1} + x^{-2} \right) dx$$

$$= \frac{1}{3} x^3 + \frac{1}{2} x^2 + x^1 + \ln|x| - x^{-1} + c$$

$$= \frac{1}{3} x^3 + \frac{1}{2} x^2 + x + \ln|x| - \frac{1}{x} + c$$

Practice

Evaluate the following integrals.

25. $\int (x^2 - 5\cos(x)) dx$

26. $\int (3e^x + 2x^3) dx$

27. $\int_e^{e^3} \frac{2}{u} du$

28. $\int (\theta + 2\sin(\theta)) d\theta$

29. $\int \left(5\sin(x) + 2e^x \right) dx$

30. $\int_0^1 (x + e^x) \, dx$

31. $\int_{\ln(2)}^{\ln(3)} 4e^x dx$

32. $\int_{\frac{\pi}{6}}^{\frac{5\pi}{2}} 8\cos(x) dx$

19 ▶ INTEGRATION BY SUBSTITUTION

The opposite of the Chain Rule is an integration technique called *substitution*. Using the Chain Rule, for example, the derivative of $8(3x^2 + 7)^5$ is $\frac{d}{dx}(8(3x^2 + 7)^5) = 8 \cdot 5(3x^2 + 7)^4 \cdot 6x = 240x(3x^2 + 7)^4$. The corresponding antiderivative is thus $\int 240x(3x^2 + 7)^4\,dx = 8(3x^2 + 7)^5 + c$. It is easy to recognize this after seeing the derivative worked out, but how should we know this otherwise?

The mantra of the Chain Rule is "multiply by the derivative of the inside." So the first step to undoing it is to identify what "the inside" must have been. We substitute a new variable u for this and then try to rewrite the whole integral in terms of u.

For example, when confronted by $\int 240x(3x^2 + 7)^4\,dx$, we first notice that if we multiplied out the fourth power, then it would be a polynomial that we know how to evaluate, but doing so would be very tiresome! Instead, we guess that "the inside" is the stuff inside the parentheses, and substitute $u = 3x^2 + 7$.

To convert the integral entirely in terms of u, we must actually get a du into the integrand. Because $u = 3x^2 + 7$, we know that $\frac{du}{dx} = 6x$, so $du = 6x\,dx$, or equivalently, $dx = \frac{du}{6x}$. So what? The steps below will illustrate why this is so useful.

Start with the original integral.

$$\int 240x(3x^2 + 7)^4\,dx$$

Substitute $u = 3x^2 + 7$ and $dx = \frac{du}{6x}$.

$$\int 240x(u)^4 \frac{du}{6x}$$

Simplify.

$$\int 40u^4\,du$$

Note, in particular, that substituting in $dx = \frac{du}{6x}$ resulted in the cancellation of all remaining x's in the integrand, so that the integral is not entirely in terms of u.

Evaluate.

$$8u^5 + c$$

Replace $u = 3x^2 + 7$.

$$8(3x^2 + 7)^5 + c$$

Thus, $\int 240x(3x^2 + 7)^4\,dx = 8(3x^2 + 7)^5 + c$, as we saw earlier.

⚠ CAUTION

Before actually computing the integral, all of the x's must be gone! You cannot integrate with mixed variables.

In general, try using something inside parentheses with u. If every x doesn't cancel out when replacing

dx with the expression involving du, then try using something else as u. Sometimes, the entire denominator can be used as u. Sometimes, nothing works and a different technique must be tried.

Example

Evaluate $\int x^2\sin(x^3)\,dx$.

Solution

If we use the stuff inside the only set of parentheses, then $u = x^3$, and thus $du = 3x^2\,dx$ and $dx = \frac{du}{3x^2}$.

Start with the original integral.

$$\int x^2\sin(x^3)\,dx$$

Substitute $u = x^3$ and $dx = \frac{du}{3x^2}$.

$$\int x^2\sin(u)\frac{du}{3x^2}$$

Simplify.

$$\int \frac{1}{3}\sin(u)\,du$$

Every x is gone, so we can evaluate.

$$-\frac{1}{3}\cos(u) + c$$

Replace $u = x^3$.

$$-\frac{1}{3}\cos(x^3) + c$$

Thus, $\int x^2\sin(x^3)\,dx = -\frac{1}{3}\cos(x^3) + c$. This can be verified by differentiating $\frac{d}{dx}\left(-\frac{1}{3}\cos(x^3) + c\right) = -\frac{1}{3}(-\sin(x^3)\cdot 3x^2) + 0 = x^2\sin(x^3)$.

If we had been faced with $\int \sin(x^3)\, dx$ in the last example, then substituting $u = x^3$ would have resulted in $\int \sin(u)\dfrac{du}{3x^2}$. This cannot be evaluated because it is not entirely in terms of u. In fact, this integral is very difficult to solve and requires the advanced technique of replacing $\sin(x^3)$ with an infinitely long polynomial called a *power series*. Many such integrals exist that are difficult to solve. This book will focus on the ones that can be evaluated with basic techniques.

Example

Evaluate $\int \dfrac{3}{2x + 7}\, dx$.

Solution

Because there are no parentheses, try using the denominator: $u = 2x + 7$. Here, $\dfrac{du}{dx} = 2$, so $du = 2\, dx$ and $dx = \dfrac{du}{2}$.

Start with the original integral.
$$\int \frac{3}{2x + 7}\, dx$$

Substitute $u = 2x + 7$ and $dx = \dfrac{du}{2}$.
$$\int \frac{3}{u}\frac{du}{2}$$

Simplify.
$$\int \frac{3}{2} \cdot \frac{1}{u}\, du$$

Every x is gone, so we can evaluate.
$$\frac{3}{2}\ln|u| + c$$

Replace $u = 2x + 7$.
$$\frac{3}{2}\ln|2x + 7| + c$$

Thus, $\int \dfrac{3}{2x + 7}\, dx = \dfrac{3}{2}\ln|2x + 7| + c$.

Basically, the goal is to find a u whose derivative is essentially the rest of the integral, except for possibly a constant, so that between the u and the du, every x goes away. This leads to some clever tricks, as will be demonstrated in the following examples.

Example

Evaluate $\int \dfrac{\ln(x)}{x}\, dx$.

Solution

Here, we use $u = \ln(x)$. This is not because it is in parentheses but because its derivative $\dfrac{du}{dx} = \dfrac{1}{x}$ makes up the rest of the integral. Here, $du = \dfrac{1}{x}\, dx$, so $dx = x\, du$.

Start with the original integral.
$$\int \frac{\ln(x)}{x}\, dx$$

Substitute $u = \ln(x)$ and $dx = x\, du$.
$$\int \frac{u}{x}(x\, du)$$

Simplify.

$$\int u\,du$$

Every x is gone, so we can evaluate.

$$\frac{1}{2}u^2 + c$$

Replace $u = \ln(x)$.

$$\frac{1}{2}(\ln(x))^2 + c$$

Thus, $\displaystyle\int \frac{\ln(x)}{x}\,dx = \frac{1}{2}(\ln(x))^2 + c$.

Example

Evaluate $\displaystyle\int \sin(x)\cos^3(x)\,dx$.

Solution

Here, the trick is to use $u = \cos(x)$ so that $\frac{du}{dx} = -\sin(x)$ and $dx = -\dfrac{du}{\sin(x)}$.

Start with the original integral.

$$\int \sin(x)\cos^3(x)\,dx$$

Substitute $u = \cos(x)$ and $dx = -\dfrac{du}{\sin(x)}$.

$$\int \sin(x) \cdot u^3\left(-\frac{du}{\sin(x)}\right)$$

Simplify.

$$\int -u^3\,du$$

Every x is gone, so we can evaluate.

$$-\frac{1}{4}u^4 + c$$

Replace $u = \cos(x)$.

$$-\frac{1}{4}\cos^4(x) + c$$

Thus, $\displaystyle\int \sin(x)\cos^3(x)\,dx = -\frac{1}{4}\cos^4(x) + c$.

To use substitution on a definite integral, evaluate the indefinite integral first, and then compute at the limits.

Example

Evaluate $\displaystyle\int_0^1 x^2 e^{x^3}\,dx$.

Solution

First, evaluate the following indefinite integral.

$$\int x^2 e^{x^3}\,dx$$

Use $u = x^3$, $du = 3x^2\,dx$, and so $dx = \dfrac{du}{3x^2}$.

$$\int x^2 e^u \frac{du}{3x^2}$$

Simplify.

$$\int \frac{1}{3}e^u\,du$$

Every x is gone, so we can evaluate the indefinite integral.

$$\frac{1}{3}e^u + c$$

Replace $u = x^3$.

$$\frac{1}{3}e^{x^3} + c$$

Now, because $\displaystyle\int x^2 e^{x^3}\,dx = \frac{1}{3}e^{x^3} + c$, it follows that

$$\int_0^1 x^2 e^{x^3}\,dx = \left[\frac{1}{3}e^{x^3}\right]_0^1 = \frac{1}{3}e^1 - \frac{1}{3}e^0 = \frac{1}{3}(e-1)$$

You will know that you chose the wrong u if either some of the variables x still remain or the simplified integral will still be hard to solve. If this happens, go back to the beginning and try a different u. Don't forget that many integrals, like those of the previous lesson, don't require substitution at all. Like much of mathematics, learning to integrate often requires patience and a knack that is developed with practice.

Practice

Evaluate the following integrals.

1. $\int x^4(x^5 + 1)^7 dx$

2. $\int (4x^4 + 3)^{10} dx$

3. $\int_0^1 x^2(x^3 - 1)^4 dx$

4. $\int (x^3 - 9x + 4) dx$

5. $\int x^3 \cdot \sqrt[3]{x^2 - 1} \, dx$

6. $\int \sqrt{2x+1} \, dx$

7. $\int_0^1 \sqrt[5]{1-x} \, dx$

8. $\int \frac{9x^2 - 5}{3x^3 - 5} dx$

9. $\int 2x^3 \cos(x^4) dx$

10. $\int \frac{6x^3 - 1}{\sqrt{3x^4 - 2x + 1}} dx$

11. $\int (8x + 5)(4x^2 + 5x - 1)^3 dx$

12. $\int \frac{x}{(4x^2 + 5)^3} dx$

13. $\int \frac{-2}{4x + 10} dx$

14. $\int \sin(x)\cos(x) dx$

15. $\int \sin^2(x)\cos(x) dx$

16. $\int \cos(4x) dx$

17. $\int 4\cos(x) dx$

18. $\int \sin(7x - 2) dx$

19. $\int e^x \sin(e^x) dx$

20. $\int \frac{(\ln(x))^3}{x} dx$

21. $\int \frac{1}{x\ln(x)} dx$

22. $\int \frac{e^{\sqrt{x}}}{\sqrt{x}} dx$

23. $\int \tan(x) dx = \int \frac{\sin(x)}{\cos(x)} dx$

24. $\int \frac{e^{2x}}{1 + e^{2x}} dx$

20 ▶ INTEGRATION BY PARTS

The integral of the product of two functions is unfortunately not the product of the integrals. For example, the integral $\int x \cdot \cos(x)dx$ is not $\left(\int x\,dx\right) \cdot \left(\int \cos(x)dx\right) = \left(\frac{1}{2}x^2\right) \cdot \left(\sin(x)\right) + c$. We know this because the derivative of $\frac{1}{2}x^2\sin(x) + c$ is, by the Product Rule, $\frac{d}{dx}\left(\frac{1}{2}x^2\sin(x) + c\right) = x \cdot \sin(x) + \cos(x) \cdot \frac{1}{2}x^2$, which is not equal to $x \cdot \cos(x)$. It is unfortunate that this does not work because, if it did, evaluating integrals would be simple and would not require so many different techniques.

The integration technique that undoes the Product Rule is called *integration by parts.* We derive the formula as follows.

The product rule for differentiating $f(x) \cdot g(x)$ says

$$\frac{d}{dx}\big(f(x) \cdot g(x)\big) = f'(x) \cdot g(x) + g'(x) \cdot f(x).$$

Integrating both sides of the formula gives $\int \frac{d}{dx}\big(f(x) \cdot g(x)\big)\,dx = \int \big[f'(x) \cdot g(x) + g'(x) \cdot f(x)\big]dx$.

This simplifies to $f(x) \cdot g(x) = \int f'(x) \cdot g(x)\,dx + \int g'(x) \cdot f(x)\,dx$.

Now, we introduce two variables, u and v, as follows to simplify the formula:

$u = f(x)$ so that $\dfrac{du}{dx} = f'(x)$, which is equivalent to $du = f'(x)dx$

$v = g(x)$ so that $\dfrac{dv}{dx} = g'(x)$, which is equivalent to $dv = g'(x)dx$

Plug these into the above formula to get $\underbrace{f(x)}_{u} \cdot \underbrace{g(x)}_{v} = \int \underbrace{g(x)}_{v} \cdot \underbrace{f'(x)dx}_{du} + \int \underbrace{f(x)}_{u} \cdot \underbrace{g'(x)\,dx}_{dv}$ preceding.

Simply put, $uv = \int v\,du + \int u\,dv$. Move the first integral on the right-hand side to the left and *voilà!*, we have the *integration by parts formula*:

$$\int u\,dv = uv - \int v\,du.$$

Work through the following examples to see how this is applied to help us compute more complicated integrals.

    This can often be used to transform a difficult integral into one that is solvable. For example, take $\int x \cdot \cos(x)dx$. This looks just like $\int u\,dv$ if $u = x$ and $dv = \cos(x)\,dx$. In order to use the formula, we will need to get du by differentiating u. Because $u = x$, we know that $\dfrac{du}{dx} = 1$, so $du = dx$. We will also need to get v from dv by integrating. And because $dv = \cos(x)dx$, it must be that $v = \sin(x)$. Thus:

$$\int x \cdot \cos(x)dx = \int u\,dv$$

$$= uv - \int v\,du$$

$$= x\sin(x) - \int \sin(x)dx$$

$$= x\sin(x) + \cos(x) + c$$

This is the correct answer, as can be verified by taking the derivative $\dfrac{d}{dx}(x\sin(x) + \cos(x) + c) = 1 \cdot \sin(x) + \cos(x) \cdot x - \sin(x) + 0 = x \cdot \cos(x)$.

⚠ CAUTION

The dx must be part of what you decide to let equal dv.

Example

Evaluate $\int xe^x dx$.

Solution

This cannot be solved by basic integration or by substitution. Since it is a product, there is a good chance that integration by parts will work. First, try $u = x$. The dv must then be everything else after the integral sign, so $dv = e^x dx$. After differentiating u and integrating dv, we get:

$$u = x$$

$$du = dx$$

And:

$$dv = e^x dx$$

$$v = e^x$$

Note: Although there is technically a "$+ c$" here, we will wait until the end to add the "$+ c$" once all integrals have been computed.

Thus, using the integration by parts formula $\int u\,dv = uv - \int v\,du$, we evaluate as follows:

$$\int xe^x\,dx = \int u\,dv$$

![⚠ CAUTION]

Don't forget that there is a *minus* in the formula!

$$= uv - \int v\,du$$

$$= xe^x - \int e^x\,dx$$

$$= xe^x - e^x + c$$

Example

Evaluate $\int x^3\ln(x)\,dx$.

Solution

Here, since we don't know how to integrate $\ln(x)$ (yet!), we can't let it equal dv. So, we set $u = \ln(x)$ and $dv = x^3\,dx$. Thus:

$$u = \ln(x)$$

$$du = \frac{1}{x}\,dx$$

And:

$$dv = x^3\,dx$$

$$v = \frac{1}{4}x^4$$

And then we evaluate.

$$\int x^3\ln(x)\,dx = \int u\,dv$$

$$= uv - \int v\,du$$

$$= \ln(x)\cdot\frac{1}{4}x^4 - \int \frac{1}{4}x^4\cdot\frac{1}{x}\,dx$$

$$= \frac{1}{4}x^4\ln(x) - \int \frac{1}{4}x^3\,dx$$

$$= \frac{1}{4}x^4\ln(x) - \frac{1}{16}x^4 + c$$

Example

Evaluate $\int \ln(x)\,dx$.

Solution

Because there seems to be only one part to this integral, one wouldn't think to try integration by parts first. However, because nothing else will work, we can try $u = \ln(x)$. The only thing left for the dv is dx, so we use $dv = dx$, which leads to $v = x$.

$$u = \ln(x)$$

$$du = \frac{1}{x}\,dx$$

And:

$$dv = dx$$

$$v = x$$

And now evaluate as follows.

$$\int \ln(x)\,dx = \int u\,dv$$

$$= uv - \int v\,du$$

$$= \ln(x)\cdot x - \int x\cdot\frac{1}{x}\,dx$$

$$= x\ln(x) - \int 1\,dx$$

$$= x\ln(x) - x + c$$

Sometimes, integration by parts needs to be done more than once to compute an integral.

Example

Evaluate $\int x^2\cos(x)\,dx$.

Solution

Here, $u = x^2$, so $du = 2x\,dx$, and $dv = \cos(x)\,dx$, so $v = \sin(x)$.

$$\int x^2\cos(x)\,dx = \int u\,dv$$

$$= uv - \int v\,du$$

$$= x^2\sin(x) - \int \sin(x)\cdot 2x\,dx$$

$$= x^2\sin(x) - \int 2x\sin(x)\,dx$$

In order to compute $\int 2x\sin(x)\,dx$, we have to use integration by parts a second time, but this time, with $u = 2x$ and $dv = \sin(x)\,dx$.

$$u = 2x$$

$$du = 2\,dx$$

And:

$$dv = \sin(x)\,dx$$

$$v = -\cos(x)$$

Now we evaluate as follows.

$$\int x^2\cos(x)\,dx = x^2\sin(x) - \int 2x\sin(x)\,dx$$

$$= x^2\sin(x) - \int u\,dv$$

$$= x^2\sin(x) - \left(uv - \int v\,du\right)$$

$$=$$

$$x^2\sin(x) - (2x\cdot(-\cos(x)) - \int(-\cos(x))\cdot 2\,dx$$

$$=$$

$$x^2\sin(x) + 2x\cos(x) + \int -2\cos(x)\,dx$$

Thus,

$$\int x^2\cos(x)\,dx =$$

$$x^2\sin(x) + 2x\cos(x) - 2\sin(x) + c$$

The final example utilizes a clever trick.

Example

Evaluate $\int e^x\sin(x)\,dx$.

Solution

The terms e^x and $\sin(x)$ are equally good candidates for u. Let us use

$$u = e^x \text{ and } dv = \sin(x)\,dx.$$

$$u = e^x$$

$$du = e^x\,dx$$

And:

$$dv = \sin(x)\,dx$$

$$v = -\cos(x)$$

Now, evaluate as follows.

$$\int e^x \sin(x)\,dx = \int u\,dv$$

$$= uv - \int v\,du$$

$$=$$

$$e^x(-\cos(x)) - \int (-\cos(x)) \cdot e^x\,dx$$

$$= -e^x\cos(x) + \int e^x\cos(x)\,dx$$

To evaluate $\int e^x\cos(x)\,dx$, we use integration by parts again, but with $u = e^x$ and $dv = \cos(x)dx$.

$$u = e^x$$

$$du = e^x dx$$

And:

$$dv = \cos(x)\,dx$$

$$v = \sin(x)$$

And then the evaluation:

$$\int e^x\sin(x)\,dx = -e^x\cos(x) + \int e^x\cos(x)\,dx$$

$$= -e^x\cos(x) + \int u\,dv$$

$$= -e^x\cos(x) + uv - \int v\,du$$

$$=$$

$$-e^x\cos(x) + e^x\sin(x) - \int \sin(x) \cdot e^x\,dx$$

Thus, we have:

$$\int e^x\sin(x)\,dx =$$

$$-e^x\cos(x) + e^x\sin(x) - \int e^x\sin(x)\,dx$$

Here is the moment of despair: To evaluate $\int e^x\sin(x)\,dx$, we need to be able to evaluate $\int e^x\sin(x)\,dx$! And yet, the trick here is to bring both integrals to one side of the equation:

$$\int e^x\sin(x)\,dx + \int e^x\sin(x)\,dx =$$
$$-e^x\cos(x) + e^x\sin(x)$$

$$2\int e^x\sin(x)\,dx = -e^x\cos(x) + e^x\sin(x)$$

$$\int e^x\sin(x)\,dx = \frac{1}{2}(-e^x\cos(x) + e^x\sin(x)) + c$$

$$= \frac{1}{2}e^x\left(-\cos(x) + \sin(x)\right) + c$$

Practice

Evaluate the following integrals using integration by parts, substitution, or basic integration.

1. $\int x^5 \ln(x)\,dx$

2. $\int x\sin(x)\,dx$

3. $\int x\sin(x^2)\,dx$

4. $\int (x + 3)\cos(x)\,dx$

5. $\int \dfrac{\ln(x)}{x}\,dx$

6. $\int x^2 \sin(x)\,dx$

7. $\int (x^2 + \sin(x))\,dx$

8. $\int x^2 e^{x^3+1}\,dx$

9. $\int \dfrac{e^{3x}}{e^{3x}+9}\,dx$

10. $\int (x^3 + 3x - 1)\ln(x)\,dx$

11. $\int \left(\dfrac{1}{x} + \ln(x)\right)dx$

12. $\int \sqrt{x - 1}\,dx$

13. $\int x\sqrt{x - 1}\,dx$

14. $\int xe^{-x}\,dx$

15. $\int \cos^3(x)\cdot\sin(x)\,dx$

16. $\int \dfrac{1}{x(\ln(x))^5}\,dx$

17. $\int \dfrac{1}{3x}\,dx$

18. $\int e^{-\cos(x)}\sin(x)\,dx$

19. $\int \dfrac{e^{\frac{1}{x}}}{x^2}\,dx$

20. $\int \sin(x)\sqrt{\cos(x)}\,dx$

21. $\int \cos(x)\cdot\ln(\sin(x))\,dx$

22. $\int e^x \cos(x)\,dx$

POSTTEST

I f you have completed all 20 lessons in this book, you are ready to take the posttest to measure your progress. The posttest has 50 multiple-choice questions covering the topics you studied in this book. Although the format of the posttest is similar to that of the pretest, the questions are different.

Take as much time as you need to complete the posttest. When you are finished, check your answers with the answer key that follows the posttest. Along with each answer is a number that tells you which lesson of this book teaches you about the calculus skills needed for that question. Once you know your score on the posttest, compare the results with the pretest. If you scored better on the posttest than you did on the pretest, congratulations! You have profited from your hard work. At this point, you should look at the questions you missed, if any. Do you know why you missed the question, or do you need to go back to the lesson and review the concept?

If your score on the posttest doesn't show much improvement, take a second look at the questions you missed. Did you miss a question because of an error you made? If you can figure out why you missed the problem, then you understand the concept and simply need to concentrate more on accuracy when taking a test. If you missed a question because you did not know how to work the problem, go back to the lesson and spend more time working that type of problem. Take the time to understand basic calculus thoroughly. You need a solid foundation in basic calculus if you plan to use this information or progress to a higher level. Whatever your score on this posttest, keep this book for review and future reference.

1. ⓐ ⓑ ⓒ ⓓ
2. ⓐ ⓑ ⓒ ⓓ
3. ⓐ ⓑ ⓒ ⓓ
4. ⓐ ⓑ ⓒ ⓓ
5. ⓐ ⓑ ⓒ ⓓ
6. ⓐ ⓑ ⓒ ⓓ
7. ⓐ ⓑ ⓒ ⓓ
8. ⓐ ⓑ ⓒ ⓓ
9. ⓐ ⓑ ⓒ ⓓ
10. ⓐ ⓑ ⓒ ⓓ
11. ⓐ ⓑ ⓒ ⓓ
12. ⓐ ⓑ ⓒ ⓓ
13. ⓐ ⓑ ⓒ ⓓ
14. ⓐ ⓑ ⓒ ⓓ
15. ⓐ ⓑ ⓒ ⓓ
16. ⓐ ⓑ ⓒ ⓓ
17. ⓐ ⓑ ⓒ ⓓ

18. ⓐ ⓑ ⓒ ⓓ
19. ⓐ ⓑ ⓒ ⓓ
20. ⓐ ⓑ ⓒ ⓓ
21. ⓐ ⓑ ⓒ ⓓ
22. ⓐ ⓑ ⓒ ⓓ
23. ⓐ ⓑ ⓒ ⓓ
24. ⓐ ⓑ ⓒ ⓓ
25. ⓐ ⓑ ⓒ ⓓ
26. ⓐ ⓑ ⓒ ⓓ
27. ⓐ ⓑ ⓒ ⓓ
28. ⓐ ⓑ ⓒ ⓓ
29. ⓐ ⓑ ⓒ ⓓ
30. ⓐ ⓑ ⓒ ⓓ
31. ⓐ ⓑ ⓒ ⓓ
32. ⓐ ⓑ ⓒ ⓓ
33. ⓐ ⓑ ⓒ ⓓ
34. ⓐ ⓑ ⓒ ⓓ

35. ⓐ ⓑ ⓒ ⓓ
36. ⓐ ⓑ ⓒ ⓓ
37. ⓐ ⓑ ⓒ ⓓ
38. ⓐ ⓑ ⓒ ⓓ
39. ⓐ ⓑ ⓒ ⓓ
40. ⓐ ⓑ ⓒ ⓓ
41. ⓐ ⓑ ⓒ ⓓ
42. ⓐ ⓑ ⓒ ⓓ
43. ⓐ ⓑ ⓒ ⓓ
44. ⓐ ⓑ ⓒ ⓓ
45. ⓐ ⓑ ⓒ ⓓ
46. ⓐ ⓑ ⓒ ⓓ
47. ⓐ ⓑ ⓒ ⓓ
48. ⓐ ⓑ ⓒ ⓓ
49. ⓐ ⓑ ⓒ ⓓ
50. ⓐ ⓑ ⓒ ⓓ

Posttest

1. Evaluate $f(-2)$ when $f(x) = x^3 - 2x$.

   a. -12

   b. -10

   c. -4

   d. 4

2. Simplify $f(2x + 1)$ when $f(x) = x^2 + x$.

   a. $4x^2 + 6x + 2$

   b. $4x^2 + 2x + 2$

   c. $2x^2 + 3x$

   d. $2x^3 + 3x^2 + x$

3. Evaluate $(g \circ h)(x)$ when $g(x) = x^2 + 5x + 1$ and $h(x) = \dfrac{1}{x}$.

   a. $x^2 + 5x + 1 + \dfrac{1}{x}$

   b. $x + 5 + \dfrac{1}{x}$

   c. $\dfrac{1}{x^2 + 5x + 1}$

   d. $\dfrac{1}{x^2} + \dfrac{5}{x} + 1$

4. What is the domain of $f(x) = \dfrac{\sqrt{x + 1}}{x}$?

   a. all real numbers greater than or equal to -1

   b. all real numbers except 0

   c. all real numbers except -1 and 0

   d. all nonzero real numbers greater than or equal to -1

Use the following graph for problems 5 through 7.

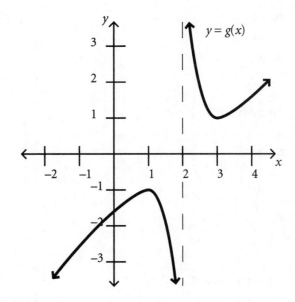

5. Where does $g(x)$ have a local maximum?

   a. $x = 0$

   b. $x = 1$

   c. $x = 2$

   d. $x = 3$

6. On what interval(s) is $g(x)$ decreasing?

   a. $(1,2)$ and $(2,3)$

   b. $(-\infty,2)$

   c. $(-\infty,0)$

   d. $(\infty,1)$ and $(3,\infty)$

7. What is the slope of the line passing through $(2,-4)$ and $(1,7)$?

a. $-\dfrac{1}{11}$

b. -11

c. 11

d. 3

8. Simplify 4^3.

a. 7

b. 12

c. 16

d. 64

9. Simplify $16^{-\frac{1}{2}}$.

a. -8

b. 4

c. $\dfrac{1}{4}$

d. $\dfrac{1}{16}$

10. Solve for x when $4^x = 10$.

a. $\dfrac{\ln(10)}{\ln(4)}$

b. 2.5

c. 10^4

d. $\ln\left(\dfrac{5}{2}\right)$

11. Evaluate $\cos\left(\dfrac{\pi}{4}\right)$.

a. 1

b. $\dfrac{1}{2}$

c. $\dfrac{\sqrt{2}}{2}$

d. $\dfrac{\sqrt{3}}{2}$

12. Evaluate $\sin\left(\dfrac{4\pi}{3}\right)$.

a. $-\dfrac{\sqrt{3}}{2}$

b. $\dfrac{\sqrt{3}}{2}$

c. $-\dfrac{\sqrt{2}}{2}$

d. $\dfrac{\sqrt{2}}{2}$

13. Evaluate $\displaystyle\lim_{x\to 2}\dfrac{x^2 - 5x + 6}{x^2 - 2x - 3}$.

a. 0

b. 1

c. -2

d. undefined

14. Evaluate $\lim\limits_{x \to 3} \dfrac{x^2 - 5x + 6}{x^2 - 2x - 3}$.

   a. -2

   b. $\dfrac{1}{4}$

   c. 1

   d. undefined

15. Evaluate $\lim\limits_{x \to 1^+} \dfrac{x + 4}{x^2 - 1}$.

   a. 5

   b. $-\infty$

   c. ∞

   d. -4

16. What is the slope of the tangent line to $f(x) = x^2$ at $x = 3$?

   a. 2

   b. 6

   c. 9

   d. $2x$

17. What is the slope of the tangent line to $y = 4x - 7$ at $x = 2$?

   a. -7

   b. -3

   c. 1

   d. 4

18. What is the derivative of $g(x) = 8x^4 - 10x^3 + 3x - 1$?

   a. $g'(x) = 0$

   b. $g'(x) = 8x^4 - 10x^3 + 3x - 1$

   c. $g'(x) = 32x^3 - 10x^2 + 3x - x$

   d. $g'(x) = 32x^3 - 30x^2 + 3$

19. Suppose that after t seconds, a falling rock is $s(t) = -16t^2 + 5t + 200$ feet off the ground. How fast is the rock traveling after 2 seconds?

   a. 10 feet per second

   b. -32 feet per second

   c. 59 feet per second

   d. 146 feet per second

20. Differentiate $y = \sqrt{x} + 4\sin(x)$.

   a. $\dfrac{dy}{dx} = \dfrac{1}{2\sqrt{x}} + 4\cos(x)$

   b. $\dfrac{dy}{dx} = \sqrt{x} + 4\cos(x)$

   c. $\dfrac{dy}{dx} = \dfrac{\sqrt{x}}{2} + 4\sin(x)$

   d. $\dfrac{dy}{dx} = \dfrac{1}{2\sqrt{x}} - 4\cos(x)$

21. What is the derivative of $f(x) = 5e^x - 2\ln(x)$?

    a. $f'(x) = 5e^x - \dfrac{2}{x}$

    b. $f'(x) = 5xe^{x-1} - \dfrac{2}{x}$

    c. $f'(x) = 5xe^{x-1} - 2x$

    d. $f'(x) = 5e^x - 2x$

22. Differentiate $y = xe^x$.

    a. $\dfrac{dy}{dx} = e^x$

    b. $\dfrac{dy}{dx} = xe^x$

    c. $\dfrac{dy}{dx} = (x + 1)e^x$

    d. $\dfrac{dy}{dx} = xe^{x-1}$

23. Differentiate $g(x) = \dfrac{\cos(x)}{x^2 + 5x}$.

    a. $g'(x) = \dfrac{\sin(x)}{2x + 5}$

    b. $g'(x) = \dfrac{-\sin(x)}{2x + 5}$

    c. $g'(x) = \dfrac{(2x + 5)\cos(x) + (x^2 + 5)\sin(x)}{(x^2 + 5)^2}$

    d. $g'(x) =$
    $\dfrac{-(x^2 + 5x)\sin(x) - (2x + 5)\cos(x)}{(x^2 + 5x)^2}$

24. What is the derivative of $f(x) = \sec(x)$?

    a. $f'(x) = \sec(x)$

    b. $f'(x) = \tan^2(x)$

    c. $f'(x) = \sec(x)\tan(x)$

    d. $f'(x) = 0$

25. What is the slope of the line that is tangent to $y = (x^2 - 2)^3$ at $x = 2$?

    a. 8

    b. 12

    c. 24

    d. 48

26. Differentiate $y = x\sin(x^2)$.

    a. $\dfrac{dy}{dx} = x\cos(x^2) + \sin(x^2)$

    b. $\dfrac{dy}{dx} = 2x^2\cos(x^2) + \sin(x^2)$

    c. $\dfrac{dy}{dx} = 2x^2\cos(x^2)$

    d. $\dfrac{dy}{dx} = -2x^2\cos(x^2) + \sin(x^2)$

27. Find $\dfrac{dy}{dx}$ when $\tan(y) + y = \ln(x) - 1$.

    a. $\dfrac{dy}{dx} = \dfrac{1}{x} - \sec^2(x)$

    b. $\dfrac{dy}{dx} = \dfrac{1}{x(1 + \sec^2(y))}$

    c. $\dfrac{dy}{dx} = \dfrac{x}{1 + \sec^2(y)}$

    d. $\dfrac{dy}{dx} = \dfrac{1}{x(1 + \sec(y)\tan(y))}$

28. Find $\dfrac{dy}{dx}$ when $x^2y = xy^2$.

   a. $\dfrac{dy}{dx} = \dfrac{1}{y}$

   b. $\dfrac{dy}{dx} = \dfrac{y - 2x}{x - 2y}$

   c. $\dfrac{dy}{dx} = \dfrac{y^2 - 2xy}{x^2 - 2xy}$

   d. $\dfrac{dy}{dx} = \dfrac{y^2 + 2xy}{x^2 + 2xy}$

29. What is the slope of the curve $y^3 - y = 3x + 3$ at $(1,2)$?

   a. $\dfrac{3}{11}$

   b. $\dfrac{3}{2}$

   c. $\dfrac{\sqrt{30}}{6}$

   d. 6

30. The volume of a sphere is $V = \dfrac{4}{3}\pi r^3$. If the radius increases by 3 meters per second, how fast is the volume changing when $r = 10$ meters?

   a. $240\pi\,\dfrac{m^3}{sec}$

   b. $\dfrac{4,000\pi m^3}{3\ sec}$

   c. $4,000\pi\,\dfrac{m^3}{sec}$

   d. $1,200\pi\,\dfrac{m^3}{sec}$

31. If a 10-foot ladder slides down a wall at 2 feet per minute (see the figure that follows), how fast does the bottom slide when the top is 6 feet up?

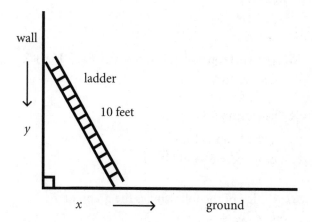

   a. $\dfrac{2}{3}$ foot per minute

   b. $\dfrac{3}{2}$ feet per minute

   c. 2 feet per minute

   d. 12 feet per minute

32. Evaluate $\displaystyle\lim_{x\to\infty} \dfrac{3x^2 + 7x - 2}{x^2 + 5x - 1}$.

   a. 2

   b. 3

   c. $\dfrac{7}{5}$

   d. ∞

33. Where does $y = e^x$ have a vertical asymptote?

   a. $y = 0$

   b. $x = 1$

   c. $x = e$

   d. no vertical asymptote

34. Evaluate $\lim\limits_{x \to \infty} \dfrac{x^5 + 3x^3}{e^x - 1}$.

   a. 5
   b. 0
   c. ∞
   d. $-3x^3$

35. On what interval(s) is $f(x) = x^3 + 6x^2 - 15x + 2$ decreasing?

   a. $(4,5)$
   b. $(-5,1)$
   c. $(2,6)$ and $(15,\infty)$
   d. $(-\infty,-5)$ and $(1,\infty)$

36. Which of the following is the graph of $y = 3x - x^3$?

a.

b.

c.

d.

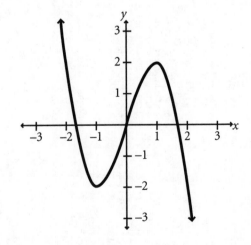

37. If up to 30 apple trees are planted on an acre, each will produce 400 apples a year. For every tree over 30 on the acre, each tree will produce 10 apples less each year. How many trees per acre will maximize the annual yield?

a. 5 trees
b. 32 trees
c. 35 trees
d. 40 trees

38. An enclosure will be built, as depicted, with 100 feet of fencing. What dimensions will maximize the area?

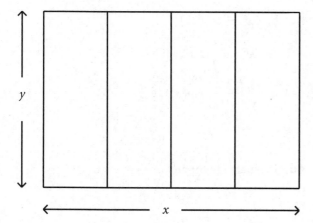

a. $x = 20$ ft, $y = 20$ ft
b. $x = 25$ ft, $y = 10$ ft
c. $x = 25$ ft, $y = 25$ ft
d. $x = 50$ ft, $y = 10$ ft

39. If $\int_{1}^{7} f(x)\,dx = 2$ and $\int_{7}^{10} f(x)\,dx = 8$, then what is $\int_{1}^{10} f(x)\,dx$?

a. 6
b. 9
c. 10
d. 16

40. What is $\int_{0}^{4} f(x)\,dx$?

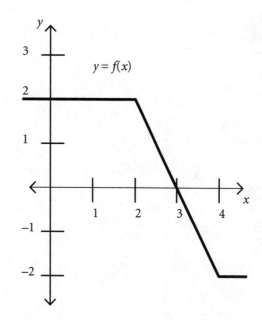

a. −4
b. 0
c. 4
d. 6

41. If $g(x) = \int_0^x t^3 \, dt$, then what is $g'(x)$?

   a. $g'(x) = x^3$

   b. $g'(x) = \frac{1}{4}t^4 + c$

   c. $g'(x) = \frac{1}{4}x^4 + c$

   d. $g'(x) = 3x^2$

42. Evaluate $\int_1^3 (6x^2 - 4x) \, dx$.

   a. 6

   b. 24

   c. 36

   d. 40

43. Evaluate $\int \left(x^5 - \frac{1}{x^2} \right) dx$.

   a. $\frac{1}{6}x^6 + \frac{3}{x^3} + c$

   b. $\frac{1}{6}x^6 + \frac{1}{x} + c$

   c. $5x^4 + \frac{2}{x^3}$

   d. $5x^4 + \frac{2}{x^3} + c$

44. Evaluate $\int \cos(x) \, dx$

   a. $-\sin(x) + c$

   b. $\sin(x) + c$

   c. $-\cos(x) + c$

   d. $\cos(x) + c$

45. Evaluate $\int (3e^x - \sin(2x)) \, dx$.

   a. $3e^x + \cos(2x) + c$

   b. $3e^x - \cos(2x) + c$

   c. $3e^x + \frac{1}{2}\cos(2x) + c$

   d. $\frac{3e^{x+1}}{x+1} - \sin(x^2) + c$

46. Evaluate $\int \frac{\ln(x)}{x} \, dx$.

   a. $\frac{1}{x} + c$

   b. $\frac{1}{2}x^2 + c$

   c. $(\ln(x))^2 + c$

   d. $\frac{1}{2}(\ln(x))^2 + c$

47. Evaluate $\int xe^{(x^2)}\,dx$.

    a. $e^{(x^2)} + c$

    b. $\dfrac{1}{2}e^{(x^2)} + c$

    c. $\dfrac{1}{2}xe^{(x^2)} + c$

    d. $\dfrac{1}{2}x^2 e^{(x^2)} + c$

48. Evaluate $\int_0^6 \sqrt{4x + 1}\,dx$.

    a. $\dfrac{2}{3}$

    b. $\dfrac{248}{3}$

    c. $\dfrac{125}{6}$

    d. 124

49. Evaluate $\int \ln(x)\,dx$.

    a. $\ln(1) + c$

    b. $\dfrac{1}{x} + c$

    c. $\dfrac{1}{2}(\ln(x))^2 + c$

    d. $x\ln(x) - x + c$

50. Evaluate $\int xe^x\,dx$.

    a. $\dfrac{1}{2}xe^x + c$

    b. $\dfrac{1}{2}x^2 e^x + c$

    c. $xe^x - e^x + c$

    d. $xe^x + e^x + c$

Answers

1. **c.** Lesson 1
2. **a.** Lesson 1
3. **d.** Lesson 1
4. **d.** Lesson 1
5. **b.** Lesson 2
6. **a.** Lesson 2
7. **b.** Lesson 2
8. **d.** Lesson 3
9. **c.** Lesson 3
10. **a.** Lesson 3
11. **c.** Lesson 4
12. **a.** Lesson 4
13. **a.** Lesson 5
14. **b.** Lesson 5
15. **c.** Lesson 5
16. **b.** Lessons 6, 7
17. **d.** Lessons 6, 7
18. **d.** Lesson 7
19. **c.** Lesson 8
20. **a.** Lesson 8
21. **a.** Lesson 8
22. **c.** Lesson 9
23. **d.** Lesson 9
24. **c.** Lesson 9
25. **d.** Lesson 10

26. **b.** Lessons 9, 10
27. **b.** Lesson 11
28. **c.** Lesson 11
29. **a.** Lesson 11
30. **d.** Lesson 12
31. **b.** Lesson 12
32. **b.** Lesson 13
33. **d.** Lesson 13
34. **b.** Lesson 13
35. **b.** Lesson 14
36. **d.** Lesson 14
37. **c.** Lesson 15
38. **b.** Lesson 15
39. **c.** Lesson 16
40. **c.** Lesson 16
41. **a.** Lesson 17
42. **c.** Lesson 18
43. **b.** Lesson 18
44. **b.** Lesson 18
45. **c.** Lessons 18, 19
46. **d.** Lesson 19
47. **b.** Lesson 19
48. **a.** Lesson 19
49. **d.** Lesson 20
50. **c.** Lesson 20

SOLUTION KEY

Lesson 1

1. $f(5) = 9$

2. $g(-3) = -20$

3. $h\left(\dfrac{1}{2}\right) = 1$

4. $f(7) = 2$. Because there is no x in the description of f, the 7 never gets used. This is called a *constant* function because it is always equal to the constant 2.

5. $m\left(-\dfrac{1}{5}\right) = -5\left(-\dfrac{1}{5}\right)^3 = -5 \cdot \left(-\dfrac{1}{5}\right)\left(-\dfrac{1}{5}\right)\left(-\dfrac{1}{5}\right)$

$= \dfrac{1}{25}$

6. $h(64) = 4$

7. The rock is $s(3) = 16$ feet high after 3 seconds.

8. The profit on 100 cookies is $P(100) = \$39$.

9. $f(y) = y^2 + 3y - 1$

10. $f(x + a) = x^2 + 2xa + a^2 + 3x + 3a - 1$

11. $\dfrac{f(x+h) - f(x)}{h} = \dfrac{\dfrac{1}{2(x+h)} - \dfrac{1}{2x}}{h} =$

$\dfrac{\dfrac{x - (x+h)}{2x(x+h)}}{h}$

$= \dfrac{x - x - h}{2hx(x+h)} = \dfrac{-h}{2hx(x+h)}$

$= \dfrac{-1}{2x(x+h)}$

12. $g(x^2 + \sqrt{x}) = \dfrac{8}{x^2 + \sqrt{x}} - 6(x^2 + \sqrt{x})$

13. $g(2x) - g(x) =$

$\dfrac{\left(-(x+a)^2 + 5\right) - \left(-x^2 + 5\right)}{a}$

14. $\dfrac{\left(-(x+a)^2 + 5\right) - \left(-x^2 + 5\right)}{a}$

$= \dfrac{\left(-(x+a)^2 + 5\right) - \left(-x^2 + 5\right)}{a}$

$= \dfrac{-2ax - a^2}{a} = \dfrac{-a(2x+a)}{a} = -(2x+a)$

15. $\dfrac{h(x+a)-h(x)}{a} = \dfrac{(-2(x+a)+1)-(-2x+1)}{a}$

$\qquad\qquad = \dfrac{-2x-2a+1+2x-1}{a}$

$\qquad\qquad = \dfrac{-2a}{a} = -2$

16. $\dfrac{g(x+2)-g(x)}{2} = \dfrac{(x+2)^3-x^3}{2}$

$\qquad = \dfrac{(x+2)\left(x^2+4x+4\right)-x^3}{2}$

$\qquad = \dfrac{x^3+6x^2+12x+8-x^3}{2} = 3x^2+6x+4$

17. $(f \circ g)(x) = \dfrac{1}{x^3-2x^2+1}$

18. $(g \circ f)(x) = \dfrac{1}{x^3} - \dfrac{2}{x^2} + 1$

19. $(f \circ h)(t) = \dfrac{1}{t-\sqrt{t}}$

20. $\left(f \circ f\right)(z) = \dfrac{1}{\frac{1}{z}} = z$

21. $\left(h \circ h\right)(w) = \left(w-\sqrt{w}\right) - \sqrt{w-\sqrt{w}}$

22. $\left(g \circ h\right)(16) = g\left(\underbrace{16-\sqrt{16}}_{=4}\right) = g(12) = 1{,}441$

23. $(h \circ f \circ g)(x) = h(f(g(x))) =$

$\qquad \dfrac{1}{x^3-2x^2+1} - \sqrt{\dfrac{1}{x^3-2x^2+1}}$

24. $(f \circ h \circ f)(2x) = f\left(h\left(\dfrac{1}{2x}\right)\right) = \dfrac{1}{\dfrac{1}{2x} - \sqrt{\dfrac{1}{2x}}}$

25. $\left(-\infty,-3\right), \left(-3,5\right),$ and $\left(5,\infty\right)$

26. $\left[-1,\infty\right)$

27. $\left(-5,\infty\right)$ (Note that t cannot equal -5.)

28. $\left(-\infty,\infty\right)$

29. $\left(-\infty,\infty\right)$

30. $\left(-\infty,\infty\right)$

31. $\left(-\infty,-8\right)$ and $\left(-8,2\right]$

32. $\left(-\dfrac{4}{3},\infty\right)$ (Note that this single interval is enough because it excludes by $-\dfrac{4}{3}$ and -3.

Lesson 2

13. There is a discontinuity at $x = 0$. The graph of f is decreasing on $(-\infty,0)$ and on $(0,\infty)$. The graph of f is concave down on $(-\infty,0)$ and concave up on $(0,\infty)$. There are no points of inflection, no local maxima, and no local minima. There is a vertical asymptote at $x = 0$ and a horizontal asymptote at $y = 0$.

14. There are no discontinuities. The function increases on $(-\infty,-3)$ and on $(0,3)$, and it decreases on $(-3,0)$ and on $(3,\infty)$. There are local maxima at $(-3,4)$ and $(3,4)$, and there is a local minimum at $(0,2)$. The graph is concave up on $(-1,1)$ and concave down on $(-\infty,-1)$ and on $(1,\infty)$. There are points of inflection at $(1,3)$ and $(-1,3)$. There are no asymptotes.

15. There is a discontinuity at $x = 1$. The function increases on $(-1,1)$ and on $(1,\infty)$, and decreases on $(-\infty,-1)$. The function is concave up on $(-\infty,1)$ and on $(1,\infty)$. There is a local minimum at $(-1,-2)$. There are no asymptotes, nor any points of inflection.

16. The discontinuities occur at $x = -2$ and $x = 2$. The function increases on $(0,1)$, $(1,2)$ and $(2,\infty)$, and decreases on $(-\infty,-2)$, $(-2,-1)$ on $(-1,0)$. The point $(0,2)$ is a local minimum. The graph is concave up on $(-2,-1)$, $(-1,1)$, and $(1,2)$ and concave down on $(-\infty,-2)$ and $(2,\infty)$. There are no points of inflection. There are vertical asymptotes at $x = -2$ and $x = 2$, and a horizontal asymptote at $y = 0$.

17. There is a discontinuity at $x = -1$. The function increases on $(-\infty,-1)$ and on $(-1,2)$, and it decreases on $(2,\infty)$. There are local maxima at $(-1,3)$ and $(2,3)$. The graph is concave up on $(-1,0)$ and concave down on $(0,\infty)$, so there is a point of inflection at $(0,2)$. Because the line is straight before $x = -1$, it does not bow upward or downward, and thus has no concavity. There are no asymptotes.

18. There are no discontinuities. The graph decreases on $(-\infty,\infty)$, is concave down on $(-\infty,0)$, and is concave up on $(0,\infty)$. There is a point of inflection at $(0,0)$. There are horizontal asymptotes at $y = -2$ and $y = 2$.

19. There are no discontinuities. The graph increases on $(0,2)$, has a local maximum at $(2,5)$, and decreases on $(2,\infty)$. The graph is concave down on $(0,3)$ and concave up on $(3,\infty)$ with a point of inflection at $(3,3)$. There is a vertical asymptote at $x = 0$ and a horizontal asymptote at $y = 1$.

20. The discontinuities occur at $x = 2$, $x = 5$, $x = 6$, and $x = 7$. The function increases on $(-\infty,1)$, $(4,5)$, and on $(5,\infty)$. The function decreases on $(1,2)$, on $(2,4)$, and on $(6,7)$. There is a local maximum at $(1,2)$ and at $(2,3)$. The points $(4,2)$ and $(17,4)$ are a local minima. The graph is concave up on $(-\infty,1)$, $(1,2)$, $(2,5)$, $(5,6)$, and $(6,7)$. Since the line is straight after $x = 7$, it does not bow upward or downward and thus has no concavity. There is a horizontal asymptote at $y = 0$ and a vertical asymptote at $x = 6$.

21. 3

22. 0

23. $-\dfrac{9}{4}$

24. $\dfrac{w - 7}{3} = \dfrac{7 - w}{-3}$

25.

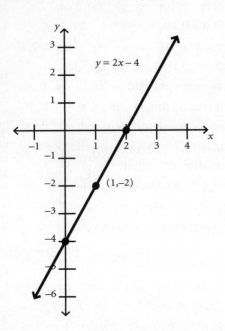

$y = 2x - 4$

$(1,-2)$

26.

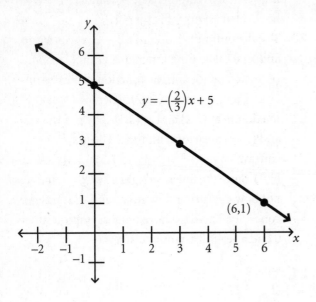

$y = -\left(\dfrac{2}{3}\right)x + 5$

$(6,1)$

27.

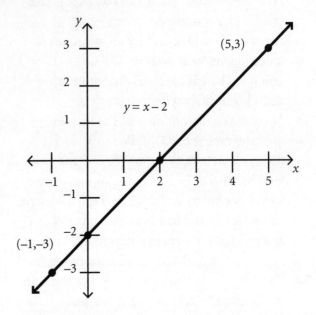

$(5,3)$

$y = x - 2$

$(-1,-3)$

28.

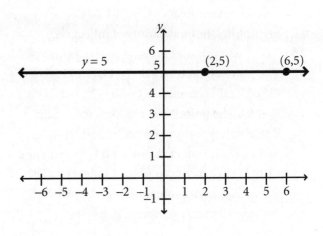

$y = 5$

$(2,5)$ $(6,5)$

Lesson 3

1. $2^5 = 32$

2. $4^3 = 64$

3. $10^4 = 10,000$

4. $6^{-2} = \dfrac{1}{36}$

5. 1

6. $3^4 = 81$

7. 9

8. $5^0 = 1$

9. $\dfrac{1}{2^3} = \dfrac{1}{8}$

10. $(3^3)^{\frac{2}{3}} = 3^{3 \cdot \frac{2}{3}} = 3^2 = 9$

11. $5^{-6} = \dfrac{1}{5^6} = \dfrac{1}{15,625}$

12. $\dfrac{8^{-2}}{8^2} = \dfrac{1}{8^4} = \dfrac{1}{4,096}$

13. $\dfrac{4^3}{4^{-1}} = 4^4 = 256$

14. $\dfrac{3^{-2}}{(3^4)^{-2}} = \dfrac{3^{-2}}{3^{-8}} = 3^6 = 729$

15. e^{11}

16. e^7

17. 1

18. 2

19. 5

20. $\ln(1) = 0$

21. $\ln(7 \cdot 2) = \ln(14)$

22. $\ln\left(\dfrac{24}{6}\right) = \ln(4)$

23. $\ln\left(\dfrac{5}{2}\right) + \ln\left(\dfrac{2}{5}\right) = \ln\left(\dfrac{5}{2} \cdot \dfrac{2}{5}\right) = \ln(1) = 0$

24. $x = \dfrac{\ln(10)}{\ln(2)}$

25. $\ln(3^x \cdot 3^5) = \ln(100)$

$\ln(3^x) + \ln(3^5) = \ln(100)$

$x\ln(3) + 5\ln(3) = \ln(100)$

$x = \dfrac{\ln(100) - 5\ln(3)}{\ln(3)} = \dfrac{\ln(100)}{\ln(3)} - 5$

26.

$\log_a\left(\dfrac{x}{y^3}\right) = \log_a x - \log_a(y^3)$

$\qquad = \log_a x - 3\log_a y = 2 - 3(-3)$

$\qquad = 2 + 9 = 11$

Lesson 4

1. $\dfrac{\pi}{6}$

2. π

3. $\dfrac{3\pi}{2}$

4. $\dfrac{5\pi}{3}$

5. $\dfrac{3\pi}{4}$

6. $60°$

7. $90°$

8. $360°$

9. $18°$

10. $330°$

11. 1

12. $\sqrt{3}$

13. 2

14. 2

15. $\dfrac{1}{\sqrt{3}} = \dfrac{\sqrt{3}}{3}$

16. $\sqrt{3}$

17. $\dfrac{2}{\sqrt{3}} = \dfrac{2\sqrt{3}}{3}$

18. $\sqrt{2}$

19. $\dfrac{\sqrt{3}}{3}$

20. $\dfrac{2\sqrt{3}}{3}$

21. $\dfrac{\sqrt{2}}{2}$

22. $-\dfrac{\sqrt{2}}{2}$

23. -1

24. $-\sqrt{2}$

25. $\sqrt{2}$

26. -1

27. 0

28. undefined

29. $\sqrt{3}$

30. $-\dfrac{\sqrt{3}}{2}$

31. $-\dfrac{1}{2}$

32. $\sqrt{3}$

33. -2

34. $-\dfrac{2\sqrt{3}}{3}$

35. $\dfrac{\sqrt{3}}{3}$

36. $\sqrt{2}$

37. $-\dfrac{1}{2}$

38. $\dfrac{\sqrt{3}}{2}$

39. $x = \dfrac{4\pi}{3}, \dfrac{5\pi}{3}$

40. $x = \pi$

Lesson 5

1. 1

2. 1

3. 1

4. undefined

5. no

6. -1

7. 4

8. undefined

9. -1

10. yes

11. 1

12. 3

13. ∞

14. ∞

15. 2

16. -2

17. 16

18. 0

19. $\dfrac{96}{23}$

20. $\dfrac{\frac{1}{2}}{\frac{\pi}{6}} = \dfrac{3}{\pi}$

21. $2x + 1$

22. $e^{-3(-2)^2} = e^{-12} = \dfrac{1}{e^{12}}$

23. $-\infty$

24. ∞

25. $\dfrac{1}{6}$

26. $-\infty$

27. $\dfrac{7}{2}$

28. $-\infty$

29. 8

30. $\lim\limits_{x\to 2} \dfrac{\cancel{x-2}}{\cancel{(x-2)}(x+2)} = \dfrac{1}{4}$

31. 1

32. $\lim\limits_{x\to 3} \dfrac{\cancel{(x-3)}(x-1)}{\cancel{(x-3)}(x+5)} = \dfrac{1}{4}$

33. $-\infty$

34. $\lim\limits_{x\to 25} \dfrac{\left(\sqrt{x}-5\right)\left(\sqrt{x}+5\right)}{(x-25)(x+1)\left(\sqrt{x}+5\right)}$

$= \lim\limits_{x\to 25} \dfrac{\cancel{(x-25)}}{\cancel{(x-25)}(x+1)\left(\sqrt{x}+5\right)} = \dfrac{1}{260}$

35. $2x$

36. $\lim\limits_{a\to 0} \dfrac{a}{\sqrt{x+a}-\sqrt{x}} \cdot \dfrac{\sqrt{x+a}+\sqrt{x}}{\sqrt{x+a}+\sqrt{x}}$

$= \lim\limits_{a\to 0} \dfrac{a\left(\sqrt{x+a}+\sqrt{x}\right)}{a}$

$= \lim\limits_{a\to 0} \left(\sqrt{x+a}+\sqrt{x}\right) = 2\sqrt{x}$

37. $\dfrac{0}{-2} = 0$

38. First note that

$$(x+h)^3 = (x+h)^2(x+h) = (x^2+2hx+h^2)(x+h)$$
$$= x^3 + 2hx^2 + h^2x + x^2h + 2h^2x + h^3$$
$$= x^3 + 3hx^2 + 3h^2x + h^3$$

So,

$$\lim\limits_{h\to 0} \dfrac{2(x+h)^3 - 2x^3}{h}$$

$$= \lim\limits_{h\to 0} \dfrac{2(x^3 + 3hx^2 + 3h^2x + h^3) - 2x^3}{h}$$

$$= \lim\limits_{h\to 0} \dfrac{2\cancel{x^3} + 6hx^2 + 6h^2x + 2h^3 - 2\cancel{x^3}}{h}$$

$$= \lim\limits_{h\to 0} \dfrac{h(6x^2 + 6hx + 2h^2)}{h}$$

$$= \lim\limits_{h\to 0}(6x^2 + 6hx + 2h^2) = 6x^2$$

39. $\lim\limits_{x\to 3} \dfrac{x^3 + 2x^2 - 15x}{x^2 - 9} = \lim\limits_{x\to 3} \dfrac{x(x-3)(x+5)}{(x-3)(x+3)}$

$$= \lim\limits_{x\to 3} \dfrac{x(x+5)}{x+3} = \dfrac{3(3+5)}{3+3}$$

$$= \dfrac{24}{6} = 4$$

40. $\lim\limits_{z\to \ln(2)} \dfrac{(e^z - 4)\cancel{(e^2 - 2)}}{-\cancel{(e^2 - 2)}} = \lim\limits_{z\to \ln(2)} -(e^z - 4)$

$$= -(e^{\ln(2)} - 4) = -(2-4)$$

$$= 2$$

Lesson 6

1. $f'(x) = \lim_{a \to 0} \dfrac{8(x + a) + 2 - (8x + 2)}{a}$

$= \lim_{a \to 0} \dfrac{8x + 8a + 2 - 8x - 2}{a} = 8$

2. $h'(x) = \lim_{a \to 0} \dfrac{(x + a)^2 + 5 - (x^2 + 5)}{a}$

$= \lim_{a \to 0} \dfrac{x^2 + 2xa + a^2 + 5 - x^2 - 5}{a}$

$= \lim_{a \to 0} 2x + a = 2x$

3. $g'(x) = \lim_{a \to 0} \dfrac{g(x + a) - g(x)}{a}$

$= \lim_{a \to 0} \dfrac{10 - 10}{a} = \lim_{a \to 0} \dfrac{0}{a} = 0$

4. $f'(x) = \lim_{a \to 0} \dfrac{3\sqrt{x + a} - 3\sqrt{x}}{a}$

$= \lim_{a \to 0} \left(\dfrac{3\sqrt{x + a} - 3\sqrt{x}}{a} \right)$

$\left(\dfrac{3\sqrt{x + a} + 3\sqrt{x}}{3\sqrt{x + a} + 3\sqrt{x}} \right)$

$= \lim_{a \to 0} \dfrac{9(x + a) - 9x}{3\sqrt{x + a} + 3\sqrt{x}}$

$= \lim_{a \to 0} \dfrac{9}{3\sqrt{x + a} + 3\sqrt{x}} = \dfrac{9}{3\sqrt{x} + 3\sqrt{x}}$

$= \dfrac{3}{2\sqrt{x}}$

5. $k'(x) = \lim_{a \to 0} \dfrac{(x + a)^3 - x^3}{a}$

$= \lim_{a \to 0} \dfrac{x^3 + 3x^2a + 3xa^2 + a^3 - x^3}{a}$

$= \lim_{a \to 0} \dfrac{(3x^2 + 3xa + a^2)a}{a}$

$= \lim_{a \to 0} (3x^2 + 3xa + a^2) = 3x^2$

6. $f'(x) = \lim_{a \to 0} \dfrac{3(x + a)^2 + (x + a) - (3x^2 + x)}{a}$

$= \lim_{a \to 0} \dfrac{6xa + 3a^2 + a}{a}$

$= \lim_{a \to 0} (6x + 3a + 1) = 6x + 1$.

Thus, at $x = 2$, the slope is $f'(2) = 13$.

7. $g'(x) = \lim_{a \to 0} \dfrac{3\sqrt{x + a} - 3\sqrt{x}}{a}$

$= \lim_{a \to 0} \dfrac{3\sqrt{x + a} - 3\sqrt{x}}{a} \cdot \dfrac{3\sqrt{x + a} + 3\sqrt{x}}{3\sqrt{x + a} - 3\sqrt{x}}$

$= \lim_{a \to 0} \dfrac{9(x + a) - 9x}{a(3\sqrt{x + a} - 3\sqrt{x})}$

$= \lim_{a \to 0} \dfrac{9a}{a(3\sqrt{x + a} - 3\sqrt{x})} = \dfrac{3}{2\sqrt{x}}$

So, the slope at $x = 16$ is $g'(16) = \dfrac{3}{8}$.

8. $g'(x) = \lim_{a \to 0} \dfrac{(x + a)^2 - 4(x + a) +}{a}$

$\dfrac{1 - (x^2 - 4x + 1)}{a}$

$= \lim_{a \to 0} \dfrac{2xa + a^2 - 4a}{a}$

$= \lim_{a \to 0} (2x + a - 4) = 2x - 4$.

Thus, there is a slope of zero when $g'(x) = 2x - 4 = 0$. This happens when $x = 2$.

9. $h'(x) = \lim_{a \to 0} \dfrac{1 - (x + a)^2 - (1 - x^2)}{a}$

$= \lim_{a \to 0} \dfrac{-2xa - a^2}{a}$

$= \lim_{a \to 0} (-2x - a) = -2x$.

The slope at $(2, -3)$ is $h'(2) = -4$, so the equation of the tangent line is
$y = -4(x - 2) - 3 = -4x + 5$.

10. $k'(x) =$

$$\lim_{a \to 0} \frac{5(x+a)^2 + 2(x+a) - (5x^2 + 2x)}{a} =$$

$$\lim_{a \to 0} \frac{10xa + 5a^2 + 2a}{a} = \lim_{a \to 0}(10x + 5a + 2) =$$

$10x + 2$. The y-value at $x = 1$ is $k(1) = 7$.
The slope at $x = 1$ is $k'(1) = 12$. The
equation of the tangent line is
$y = 12(x - 1) + 7 = 12x - 5$.

Lesson 7

1. $f'(x) = 5x^4$

2. $\dfrac{dy}{dx} = 21x^{20}$

3. $g'(u) = -5u^{-6} = \dfrac{-5}{u^6}$

4. $h'(x) = 0$

5. $\dfrac{dy}{dt} = 12t^{11}$

6. $\dfrac{dy}{dx} = \dfrac{7}{5} \cdot x^{\frac{2}{5}}$

7. $f'(x) = 100x^{99}$

8. $f'(t) = 0$

9. $g'(x) = -\dfrac{4}{5}x^{-\frac{9}{5}} = -\dfrac{4}{5x^{\frac{9}{5}}}$

10. $k(x) = x^{\frac{1}{4}}$, so $k'(x) = \dfrac{1}{4}x^{-\frac{3}{4}} = \dfrac{1}{4x^{\frac{3}{4}}}$

11. $y = u^{\frac{1}{2}}$, so $\dfrac{dy}{du} = \dfrac{1}{2}u^{-\frac{1}{2}} = \dfrac{1}{2\sqrt{u}}$

12. $y = x^{-1}$, so $\dfrac{dy}{dx} = -x^{-2} = -\dfrac{1}{x^2}$

13. $f(x) = \dfrac{1}{x^{\frac{1}{2}}} = x^{-\frac{1}{2}}$, so $f'(x) = -\dfrac{1}{2}x^{-\frac{3}{2}} = -\dfrac{1}{2x^{\frac{3}{2}}}$

14. $g(x) = \dfrac{1}{x \cdot x^{\frac{1}{2}}} = \dfrac{1}{x^{\frac{3}{2}}} = x^{-\frac{3}{2}}$.

So, $g'(x) = -\dfrac{3}{2}x^{-\frac{5}{2}} = -\dfrac{3}{2x^{\frac{5}{2}}}$.

15. $h(t) = t^7$. So, $h'(t) = 7t^6$.

16. $y = \dfrac{t^{\frac{1}{3}}}{t^{\frac{1}{4}}} = t^{\frac{1}{3} - \frac{1}{4}} = t^{\frac{1}{12}}$. So, $\dfrac{dy}{dt} = \dfrac{1}{12}t^{-\frac{11}{12}} = \dfrac{1}{12t^{\frac{11}{12}}}$.

17. $\dfrac{dy}{dx} = 42x^6$

18. $f'(x) = 30x^{-11} = \dfrac{30}{x^{11}}$

19. $V'(r) = 4\pi r^2$

20. $g'(t) = \dfrac{48}{5}t^3$

21. $k'(x) = -2x$

22. $\dfrac{dy}{dt} = 12t^2 - 6t$

23. $f'(x) = 24x^2 + 6x$

24. $\dfrac{dy}{dx} = -4x^{-3} - 2(-3x^{-4}) - 1 = -\dfrac{4}{x^3} + \dfrac{6}{x^4} - 1$

25. $s'(t) = 3\pi t^2 + 2et + \sqrt{3}$

26. $F'(x) = 600x^{99} + 500x^{49} - 100x^{24} + 20x^9$

27. $g'(x) = \dfrac{3}{5}x^{-\frac{4}{5}} + 15x^2 = \dfrac{3}{5x^{\frac{4}{5}}} + 15x^2$

28. $h'(u) = 5u^4 + 16u^3 - 21u^2 - 4u + 8$

29. $\dfrac{dy}{dx} = -2x^{-2} - 2x^{-3} = -\dfrac{2}{x^2} - \dfrac{2}{x^3}$

30. $\dfrac{dy}{du} = 2u + 2u^{-3} = 2u + \dfrac{2}{u^3}$

31. $\dfrac{dy}{dx} = 2x^{-\frac{1}{2}} + 3x^{-\frac{2}{3}} = \dfrac{2}{\sqrt{x}} + \dfrac{3}{x^{\frac{2}{3}}}$

32. $f(x) = \dfrac{3}{2}x^4 - 5x^{-\frac{3}{2}}$, so $f'(x) = 6x^3 + \dfrac{15}{2}x^{-\frac{5}{2}}$

33. $f'(x) = -x^{-2}$, $f''(x) = 2x^{-3}$,
$f'''(x) = -6x^{-4}$,
and $f''''(x) = 24x^{-5}$

34. $s''(t) = -32$

35. $\dfrac{d^3y}{dx^3} = -180x^{-6} + 48x^{-5} - 6x^{-4}$

36. $\dfrac{dy}{dt} = 2t^{-\frac{2}{3}}$, $\dfrac{d^2y}{dt^2} = -\dfrac{4}{3}t^{-\frac{5}{3}}$, and $\dfrac{d^3y}{dt^3} = \dfrac{20}{9}t^{-\frac{8}{3}}$

Lesson 8

1. pay rate in dollars per hour
2. fuel economy in miles per gallon
3. baby's growth rate in pounds per month
4. rate at which the radius shrinks in inches per hour

5. $h'(t) = \dfrac{25}{t^2}$, so $h'(5) = 1$. So, it is increasing at the rate of 1 foot per year.

6. $L'(t) = -2t + 8$, so $L'(7) = -6$. So, it is decreasing at the rate of 6 feet per day.

7. $P'(x) = \dfrac{3}{10}x^2 - 120x + 9,000$, so $P'(50) = 3,750$. So, the profit is increasing at the rate of \$3,750 per car sold.

8. $C'(3) = 4.8 - \dfrac{8}{3} \approx 2.13$, so the cost would increase at a rate of \$2.13 per inch at the instant when $x = 3$ inches.

9. After 3 seconds, it is at $s(3) = 75$ meters from the start. Note that $v(t) = 3t^2 + 4t + 10$ and $a(t) = 6t + 4$. At that moment, it is traveling at $v(3) = 49$ meters per second and accelerating at $a(3) = 22$ meters per second per second.

10. The position function is $s(t) = -16t^2 + 64$, the velocity function is $v(t) = -32t$, and the acceleration is a constant $a(t) = -32$. Since $s(2) = 0$, it will hit the ground when $t = 2$ seconds and be traveling at $v(2) = -64$ feet per second (downward) at that instant.

11. The position function is $s(t) = -16t^2 + 800t$ and the velocity function is $v(t) = -32t + 800$. The bullet will stop in the air when the velocity is zero. This happens at $t = 25$ seconds, when the bullet is $s(25) = 10,000$ feet in the air.

12. The position function is $s(t) = -16t^2 - 10t + 1,000$, so after 4 seconds, it is $s(4) = 704$ feet off the ground. It has therefore fallen 296 feet by this moment. Since $v(t) = -32t - 10$, it is traveling $v(4) = -138$ feet per second (downward) at this moment.

13. $\dfrac{dy}{dx} = 20x^4 - 10\sin(x)$

14. $f'(t) = 3\cos(t) - \dfrac{2}{t^2}$

15. $g'(x) = -6x^{-3} - \dfrac{5}{3}x^{-\frac{4}{3}} + \pi\sin(x)$
$= -\dfrac{6}{x^3} - \dfrac{5}{3x^{\frac{4}{3}}} + \pi\sin(x)$

16. $r'(\theta) = \dfrac{1}{2}\cos(\theta) - \dfrac{1}{2}\sin(\theta)$

17. $h'(x) = -\sin(x)$ because $\cos(5)$ is a constant

18. Because $f\left(\dfrac{\pi}{2}\right) = \sin\left(\dfrac{\pi}{2}\right) + \cos\left(\dfrac{\pi}{2}\right) = 1 + 0 = 1$, the point is $\left(\dfrac{\pi}{2}, 1\right)$. Since $f'(x) = \cos(x) - \sin(x)$, the slope is $f'\left(\dfrac{\pi}{2}\right) = -1$, so the equation is $y = -\left(x - \dfrac{\pi}{2}\right) + 1 = -x + \dfrac{\pi}{2} + 1$.

19. $f'(x) = 1 + 2x + 3x^2 + e^x$

20. $g'(t) = \dfrac{12}{t} + 2t$

21. $\dfrac{dy}{dx} = -\sin(x) - 10e^x + 8$

22. $h'(x) = \dfrac{1}{2\sqrt{x}} - \dfrac{8}{x}$. (Note that e^{-3} is a constant.)

23. $k'(u) = \dfrac{15}{2}x^{\frac{3}{2}} + 5e^x$. (Note that $\ln(\pi)$ is a constant.)

24. $f'(x) = e^x + \dfrac{1}{x}$, so $f''(x) = e^x - \dfrac{1}{x^2}$

25. $g^{(100)}(x) = 3e^x$

26. $f'(10) = \dfrac{1}{10}$

Lesson 9

1. $f'(x) = 2x\cos(x) - \sin(x) \cdot x^2 = x(2\cos(x) - x\sin(x))$

2. $\dfrac{dy}{dt} = 24t^2 e^t + e^t \cdot 8t^3 = 8t^2 e^t(3 + t)$

3. $\dfrac{dy}{dx} = \pi(\cos^2(x) - \sin^2(x))$

4. $g'(x) = 6x\ln(x) + \dfrac{1}{x}(3x^2) - 20x^3$
$= x(6\ln(x) + 3 - 20x^2)$

5. $h'(u) = e^u(u^2 + 3u) + (2u + 3)e^u$
$= e^u(u^2 + 5u + 3)$

6. $k'(x) = \dfrac{1}{2}x^{-\frac{1}{2}} \cdot \cos(x) - \sin(x) \cdot x^{\frac{1}{2}}$
$- \left(\cos(x) \cdot x^{-\frac{1}{4}} - \dfrac{1}{4}x^{-\frac{5}{4}} \cdot \sin(x)\right)$

7. $\dfrac{dy}{dx} = \dfrac{8\sin(x)}{x} + 8\ln(x)\cos(x) - \sin(x)$

8. $h'(t)$
$= 3t^2\left(\sin(t) - \cos(t)\right) + \left(\cos(t) + \sin(t)\right)\left(t^2 + 4\right)$

9. $\dfrac{dy}{dx} = 15x^2 - \left(1 \cdot \ln(x) + \dfrac{1}{x} \cdot x\right)$
$= 15x^2 - \ln(x) - 1$

10. $f'(x) = \cos(x)\sin(x) + \cos(x)\sin(x)$
$= 2\cos(x)\sin(x)$

11. $\dfrac{dy}{dx} = e^x\sin(x) + (e^x\sin(x) + \cos(x)e^x) \cdot x$

12. $g'(x) = 12x^3\ln(x)\cos(x) +$
$\left(\dfrac{1}{x}\cos(x) - \sin(x)\ln(x)\right) \cdot 3x^4$
$= 12x^3\ln(x)\cos(x) +$
$3x^3\cos(x) - 3x^4\sin(x)\ln(x)$

13. $f'(x) = 2xe^x + e^x x^2 + 1$.
So, $f'(0) = 2(0)e^0 + e^0(0)^2 + 1 = 1$,
so the slope is 1.

14. $y = -\pi(x - \pi) = -\pi x + \pi^2$

15. $h'(x) = \dfrac{(3x^2 + 10)(3x^2 + 5x + 2)}{(3x^2 + 5x + 2)^2} -$
$\dfrac{(6x + 5)(x^3 + 10x + 7)}{(3x^2 + 5x + 2)^2}$

16. $f'(x) =$
$\dfrac{2x(x^2 + 1) - 2x(x^2 - 1)}{(x^2 + 1)^2} = \dfrac{4x}{(x^2 + 1)^2}$

17. $f'(x) = \dfrac{(1 + \frac{1}{x})(e^x - 1) - e^x(x + \ln(x))}{(e^x - 1)^2}$

18. $\dfrac{dy}{dx} = \dfrac{5x^4\ln(x) - \frac{1}{x} \cdot x^5}{(\ln(x))^2} = \dfrac{5x^4\ln(x) - x^4}{(\ln(x))^2}$
$\dfrac{x^4\left(5\ln(x) - 1\right)}{\left(\ln(x)\right)^2}$

19. $\dfrac{dy}{dt} =$
$\dfrac{(4e^t + 1)(t^3 + 2t + 1) - (3t^2 + 2)(4e^t + t)}{(t^3 + 2t + 1)^2}$

20. $g't = \dfrac{3\pi t^2 \sin(t) - \pi t^3 \cos(t)}{\pi^2 \sin^2(t)}$

$= \dfrac{3t^2 \sin(t) - t^3 \cos(t)}{\pi \sin^2(t)}$

21. $\dfrac{dy}{dx} = \dfrac{-\dfrac{3}{2}x^{\frac{1}{2}}\left(1+x^{\frac{3}{2}}\right) - \dfrac{3}{2}x^{\frac{1}{2}}\left(1-x^{\frac{3}{2}}\right)}{\left(1+x^{\frac{3}{2}}\right)^2}$

$= \dfrac{-3x^{\frac{1}{2}}}{\left(1+x^{\frac{3}{2}}\right)^2}$

22. $g'(u) = \dfrac{\cos(u)\cdot(u^3 - 3u) - (3u^2 - 2)\cdot\sin(u)}{\left(u^3 - 3u\right)^2}$

23. $\dfrac{dy}{dx}$

$= \dfrac{0 - \left(\cos(x)\cdot(\cos(x)-\pi) - \sin(x)\cdot(\sin(x)+\pi)\right)}{\left(\sin(x)+\pi\right)^2 \left(\cos(x)-\pi\right)^2}$

24. $h'(t) =$

$\dfrac{\left(\dfrac{1}{t}+1\right)\cdot\sin^2(t) - \left(\overbrace{\cos(t)\cdot\sin(t) + \cos(t)\cdot\sin(t)}^{=2\sin(t)\cos(t)}\right)\cdot\left(\ln(t)+t\right)}{\sin^4(t)}$

25. $\dfrac{dy}{dx} = \dfrac{(\ln(x)+1)e^x - xe^x\ln(x)}{e^x \cdot e^x}$

$= \dfrac{\ln(x) + 1 - x\ln(x)}{e^x}$

26. $f'(x) = \dfrac{(2xe^x + e^x \cdot x^2)\cos(x) + \sin(x)\cdot x^2 e^x}{\cos^2(x)}$

27. $\dfrac{dy}{dx} = 1\cdot e^x + e^x \cdot x + e = e^x\cdot(x+1) + e$

$\dfrac{d^2 y}{dx^2} = e^x\cdot(x+1) + 1\cdot e^x = e^x\cdot(x+2)$

28. $f'(x) = 2x\cdot\ln(x) + x$, so

$f'(e) = 2e\underbrace{\ln(e)}_{=1} + e = 2e + e = 3e$.

29. $\dfrac{d}{dx}(\csc(x)) = \dfrac{d}{dx}\left(\dfrac{1}{\sin(x)}\right)$

$= -\dfrac{\cos(x)}{\sin^2(x)} = -\dfrac{1}{\sin(x)}\cdot\dfrac{\cos(x)}{\sin(x)}$

$= -\csc(x)\cot(x)$

30. $\dfrac{d}{dx}(\cot(x)) = \dfrac{d}{dx}\left(\dfrac{\cos(x)}{\sin(x)}\right)$

$= \dfrac{-\sin^2(x) - \cos^2(x)}{\sin^2(x)}$

$= \dfrac{-1}{\sin^2(x)} = -\csc^2(x)$

31. $f'(x) = \tan(x) + x\cdot\sec^2(x)$

32. $g'(x) = e^x \sec(x) + \sec(x)\tan(x)e^x$

$= e^x \sec(x)(1+\tan(x))$

33. $h'(t) = \left(e^t\cdot\ln(t) + \dfrac{1}{t}\cdot e^t\right)\tan(t) + \sec^2(t)$

$\cdot e^t\cdot\ln(t)$

34. $j'(x) =$

$\dfrac{\left(\dfrac{1}{3}x^{-\frac{2}{3}} + \dfrac{1}{4}x^{-\frac{3}{4}}\right)\sec(x) - \sec(x)\tan(x)\cdot\left(x^{\frac{1}{3}} + x^{\frac{1}{4}}\right)}{\sec^2(x)}$

Lesson 10

1. $f'(x) = 4(8x^3 + 7)^3\cdot(24x^2)$

2. $\dfrac{dy}{dx} = 3(x^2 + 8x + 9)^2\cdot(2x + 8)$

3. $h'(t) = 10(t^8 - 9t^3 + 3t + 2)\cdot$
$(8t^7 - 27t^2 + 3)$

4. $\dfrac{dy}{du} = \dfrac{7}{2}(u^5 - 3u^4 + 7)^{\frac{5}{2}}\cdot(5u^4 - 12u^3)$

5. $g'(x) = \dfrac{1}{2}(x^2 + 9x + 1)^{-\frac{1}{2}}\cdot(2x + 9)$

$= \dfrac{2x + 9}{2\sqrt{x^2 + 9x + 1}}$

6. $\dfrac{dy}{dx} = \dfrac{1}{3}\left(e^x + 1\right)^{-\frac{2}{3}} \cdot e^x = \dfrac{e^x}{3\left(e^x + 1\right)^{\frac{2}{3}}}$

7. $f'(x) = \sec^2\left(\sqrt{x}\right) \cdot \dfrac{1}{2\sqrt{x}}$

8. $g'(x) = \dfrac{1}{2}\left(\tan(x)\right)^{-\frac{1}{2}} \cdot \sec^2(x)$

9. $\dfrac{dy}{dt} = \dfrac{3}{3t + 5}$

10. $h'(x) = \pi\cos(\pi x)$

11. $\dfrac{dy}{dx} = \dfrac{5(\ln(x))^4}{x}$

12. $f'(x) = e^x + 2e^{2x} + 3e^{3x}$

13. $g(x) = \dfrac{1}{2}\left(e^x - e^{-x}\right)$, so $g'(x) = \dfrac{1}{2}\left(e^x + e^{-x}\right)$

14. $f'(\theta) = \dfrac{\cos(2\theta) \cdot 2 \cdot \theta - 1 \cdot \sin(2\theta)}{\theta^2}$

$= \dfrac{2\theta\cos(2\theta) - \sin(2\theta)}{\theta^2}$

15. $\dfrac{dy}{dx} = e^{2x} + 2xe^{2x}$

16. $f'(x) =$
$\sec(10x^2 + e^x)\tan(10x^2 + e^x) \cdot (20x + e^x)$

17. $\dfrac{dy}{dx} = e^{x^2} \cdot 2x + 2ex^{2e-1}$

18. $\dfrac{dy}{dx} = \dfrac{1}{x\sin(x)} \cdot \left(\sin(x) + \cos(x) \cdot x\right)$

19. $s'(u) = 3\left(\sin(u) + \cos(u)\right)^2 \cdot \left(\cos(u) - \sin(u)\right)$

20. $\dfrac{dy}{dx} = \sec^2\left(\cos(x)\right) \cdot \left(-\sin(x)\right)$

21. $f'(x) = 3\cos^2(8x) \cdot \left(-\sin(8x)\right) \cdot 8$

$= -24\cos^2(8x)\sin(8x)$

22. $\dfrac{dy}{dx} = 4\left(e^{9x^2 + 2x + 1}\right)^3 \cdot \left(e^{9x^2 + 2x + 1}\right) \cdot (18x + 2)$

23. $g'(t) = \dfrac{\sec^2(e^t + 1) \cdot e^t}{\tan(e^t + 1)}$

24. $\dfrac{dy}{dx} = \cos(\sin(\sin(x))) \cdot \cos(\sin(x)) \cdot \cos(x)$

25. $k'(u) = \sec(\ln(8u^3))\tan(\ln(8u^3)) \cdot \dfrac{1}{8u^3} \cdot 24u^2$

$= \dfrac{3}{u}\sec(\ln(8u^3))\tan(\ln(8u^3))$

26. $h'(x) = 2\cos\left(\dfrac{e^{4x^2} - 1}{1 - e^x}\right) \cdot \left(-\sin\left(\dfrac{e^{4x^2} - 1}{1 - e^x}\right)\right)$

$\cdot \left(\dfrac{e^{4x^2} \cdot 8x \cdot \left(1 - e^x\right) + e^x \cdot \left(e^{4x^2} - 1\right)}{\left(1 - e^x\right)^2}\right)$

Lesson 11

1. $3(y + 1)^2 \cdot \dfrac{dy}{dx} = 4x^3 - 8$, so

$\dfrac{dy}{dx} = \dfrac{4x^3 - 8}{3(y + 1)^2}$

2. $3y^2 \cdot \dfrac{dy}{dx} + \dfrac{dy}{dx} = \cos(x)$, so $\dfrac{dy}{dx} = \dfrac{\cos(x)}{3y^2 + 1}$

3. $\dfrac{dy}{dx} = \dfrac{4}{\cos(y)} = 4\sec(y)$

4. $\dfrac{dy}{dx} = \dfrac{\frac{1}{x}}{1 - \frac{1}{2\sqrt{y}}} = \dfrac{2\sqrt{y}}{2x\sqrt{y} - x}$

5. $\dfrac{dy}{dx} = \dfrac{12x^3 - 1}{2y - 8}$

6. $\dfrac{dy}{dx} = \dfrac{e^x + 2e^{2x}}{e^y + 2e^{2y}}$

7. $\dfrac{dy}{dx} = \dfrac{-\sin(x)}{\sec^2(y)} = -\sin(x)\cos^2(y)$

8. $\dfrac{dy}{dx} = \dfrac{1}{2\sqrt{x+y}}\left(1 + \dfrac{dy}{dx}\right)$, so

$$\dfrac{dy}{dx} = \dfrac{\dfrac{1}{2\sqrt{x+y}}}{1 - \dfrac{1}{2\sqrt{x+y}}} = \dfrac{1}{2\sqrt{x+y}-1}$$

9. $\dfrac{dy}{dx} = \dfrac{-\cos(x)+\cos(x-y)}{-\cos(y)+\cos(x-y)}$

> ### ⚠ CAUTION
> Though tempting, don't cancel the terms $\cos(x-y)$.

10. $\dfrac{dy}{dx} = \dfrac{-1}{x(3y^2-2y-2)}$

11. $\dfrac{dy}{dx} = \dfrac{5}{2(y+x^2)^3} - 2x$

12. $\dfrac{dy}{dx} = \dfrac{4x^3y^4-2xy}{x^2-4y^3x^4}$

13. $\dfrac{y - x\cdot\frac{dy}{dx}}{y^2} + y + \dfrac{dy}{dx}\cdot x = 1 + \dfrac{dy}{dx}$, so

$$\dfrac{dy}{dx} = \dfrac{1-y-\dfrac{1}{y}}{-\dfrac{x}{y^2}+x-1} = \dfrac{y^2-y^3-y}{-x+xy^2-y^2}$$

14. $\dfrac{dy}{dx} = \dfrac{3x^2\cos(y)}{\sec(y)\tan(y)+9+x^3\sin(y)}$

15. $3y^2\cdot\dfrac{dy}{dx} + 2x = 2y\cdot\dfrac{dy}{dx} - 5\cdot\dfrac{dy}{dx}$, so at

$(-3,1)$, the tangent slope is $\dfrac{dy}{dx} = 1$.

16. $\dfrac{dy}{dx} = \dfrac{-3x^2-1}{3y^2-3}$, so at $(1,-2)$ the tangent slope

is $-\dfrac{4}{9}$.

17. $\dfrac{dy}{dx} = \dfrac{-1}{\dfrac{3}{3y-5}-2y}$, so at $(4,2)$ the tangent slope is 1.

18. $\dfrac{dy}{dx} = \dfrac{-1}{\dfrac{\pi}{2\sqrt{2}}\cos(y)\cdot\sin\left(\dfrac{\pi}{2\sqrt{2}}\sin(y)\right)}$, so at

$\left(\dfrac{\sqrt{2}}{2},\dfrac{\pi}{4}\right)$ the tangent slope is $-\dfrac{4\sqrt{2}}{\pi}$.

19. $\dfrac{dy}{dx} = \dfrac{2\sqrt{3}}{3}$ at $\left(\dfrac{1}{2},\dfrac{\pi}{6}\right)$, so the tangent

equation is $y = \dfrac{2\sqrt{3}}{3}\left(x-\dfrac{1}{2}\right) + \dfrac{\pi}{6}$.

20. $\dfrac{dy}{dx} = \dfrac{-\dfrac{e^x}{e^x+e^y}}{\dfrac{e^x}{e^x+e^y}+1}$, so at $(0,0)$ the tangent slope is

$-\dfrac{1}{3}$. As such, the equation of the tangent line is

$y = -\dfrac{1}{3}x$.

Lesson 12

1. $\dfrac{dy}{dt} = 5(x^3+x-1)^4\cdot\left(3x^2\dfrac{dx}{dt}+\dfrac{dx}{dt}\right)$

2. $4y^3\dfrac{dy}{dt} - 6x\dfrac{dx}{dt} = -\sin(y)\dfrac{dy}{dt}$

3. $\dfrac{1}{2\sqrt{x}}\cdot\dfrac{dx}{dt} + \dfrac{1}{2\sqrt{y}}\cdot\dfrac{dy}{dt} = 30x^2\cdot\dfrac{dx}{dt} - 7\cdot\dfrac{dx}{dt}$

4. $\dfrac{1}{y}\cdot\dfrac{dy}{dt} + e^x\cdot\dfrac{dx}{dt} = 2x\cdot\dfrac{dx}{dt}\cdot y^2 + 2y\cdot\dfrac{dy}{dt}\cdot x^2$

5. $\dfrac{dz}{dt} = \dfrac{4}{5}x\cdot\dfrac{dx}{dt} + \dfrac{4}{5}y\cdot\dfrac{dy}{dt} - \dfrac{3}{5x^2}\cdot\dfrac{dx}{dt}$

6. $2A\cdot\dfrac{dA}{dt} + 2B\cdot\dfrac{dB}{dt} = 2C\cdot\dfrac{dC}{dt}$

7. $\dfrac{dV}{dt} = 4\pi r^2 \cdot \dfrac{dr}{dt}$

8. $\dfrac{dA}{dt} = 8\pi r \cdot \dfrac{dr}{dt}$

9. $\dfrac{dC}{dt} = 2\pi \cdot \dfrac{dr}{dt}$

10. $\dfrac{dA}{dt} = \dfrac{1}{2} \cdot \dfrac{db}{dt} \cdot h + \dfrac{dh}{dt} \cdot \dfrac{1}{2}b$

$\quad = \dfrac{1}{2}\left(\dfrac{db}{dt} \cdot h + \dfrac{dh}{dt} \cdot b\right)$

11. $\dfrac{dS}{dt} = 12e \cdot \dfrac{de}{dt}$

12. $\dfrac{dD}{dt} = \dfrac{1}{2\sqrt{x^2 + y^2}} \cdot \left(2x \cdot \dfrac{dx}{dt} + 2y \cdot \dfrac{dy}{dt}\right)$

$\quad = \dfrac{x \cdot \dfrac{dx}{dt} + y \cdot \dfrac{dy}{dt}}{\sqrt{x^2 + y^2}}$

13. $\dfrac{dx}{dt} = -\dfrac{35}{12}$

14. $\dfrac{dy}{dt} = -\dfrac{4}{3}$

15. $\dfrac{dK}{dt} = 24$

16. $\dfrac{dB}{dt} = 28$

17. Because $\dfrac{dA}{dt} = 172$, A is increasing at the rate of 172 square feet per minute when $I = 20$.

18. $\dfrac{dR}{dt} = \dfrac{27}{64}$, so R is increasing at the rate of $\dfrac{27}{64}$ per hour at this instant.

19. $\dfrac{dA}{dt} = -65$, so the area is decreasing at the rate of 65 square feet per minute.

20. $\dfrac{dA}{dt} = -320\pi \dfrac{in^2}{hour}$, so the area shrinks by 320π square inches per hour.

21. $\dfrac{dr}{dt} = \dfrac{5}{2\pi}$, so the radius grows at $\dfrac{5}{2\pi} \approx 0.796$ feet per hour.

22. $\dfrac{ds}{dt} = 4$, so each side is shrinking at the rate of 4 inches per minute.

23. $\dfrac{de}{dt} = -\dfrac{1}{240}$, so each side is growing at the rate of $\dfrac{1}{240}$ inch per second.

24. $\dfrac{db}{dt} = 7$, so the base increases at the rate of 7 inches per hour.

25. If the height is y and the base is x, then $x^2 + y^2 = 10^2$ and $2x\dfrac{dx}{dt} + 2y\dfrac{dy}{dt} = 0$. After 6 seconds, $y = 6$ and $x^2 + 6^2 = 100$, so $x = 8$. Because $\dfrac{dy}{dt} = 1, 2(8)\dfrac{dx}{dt} + 2(6)(1) = 0$, so $\dfrac{dx}{dt} = -\dfrac{3}{4}$. The end of the board is moving at the rate of $\dfrac{3}{4}$ of a foot each hour along the ground.

26. If the base is x and the hypotenuse (length of the string) is s, then $x^2 + 100^2 = s^2$. Using this, when $s = 260$, x must be 240. Because $\dfrac{dx}{dt} = 13$, we can calculate that $\dfrac{ds}{dt} = 12$. Thus, the string is being let out at 12 feet per second.

Lesson 13

1. 0

2. $\dfrac{4}{5}$

3. $\dfrac{5}{2}$

4. ∞

5. 0

6. $-\dfrac{8}{9}$

7. $-\infty$

8. ∞

9. 1

10. 0

11. 3

12. ∞

13. $-\infty$

14. ∞

15. 0

16. 0

17. ∞

18. 0

19. 0

20. $-\dfrac{4}{7}$

21. vertical asymptote at $x = 4$, horizontal asymptote at $y = 1$, sign diagram:

22. vertical asymptotes at $x = 2$ and $x = -2$,

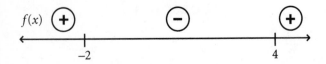

horizontal asymptote at $y = 0$, sign diagram:

23. vertical asymptote at $x = -3$, horizontal asymptote at $y = 1$, sign diagram:

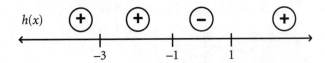

24. vertical asymptotes at $x = 1$ and $x = 3$, horizontal asymptote at $y = 0$, sign diagram:

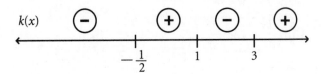

25. vertical asymptote at $x = -5$, horizontal asymptote at $y = 0$, sign diagram:

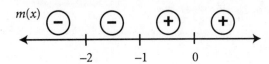

26. vertical asymptotes at $x = -2$ and $x = -1$, no horizontal asymptote, sign diagram:

27. ∞

28. ∞

29. ∞

30. $-\infty$

31. $-\infty$

32. $-\infty$

Lesson 14

1. $f(x)$ has no asymptotes. $f'(x) = 2x - 30$, thus there is a local minimum at $(15, -215)$. Because $f''(x) = 2$, the graph is always concave up.

2. $g(x)$ has no asymptotes.
$g'(x) = -4 - 2x = -2(2 + x)$, so there is a local maximum at $(-2, 4)$. Because $g''(x) = -2$, the graph is always concave down.

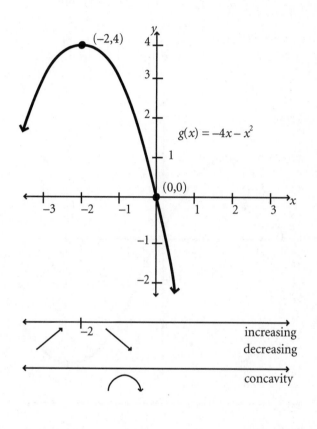

3. $h(x)$ has no asymptotes.

$h'(x) = 6x^2 - 6x - 36 = 6(x - 3)(x + 2)$, so there is a local maximum at $(-2, 49)$ and a local minimum at $(3, -76)$. Because $h''(x) = 12x - 6 = 12\left(x - \dfrac{1}{2}\right)$, there is a point of inflection at $\left(\dfrac{1}{2}, \dfrac{-27}{2}\right)$.

4. $k(x)$ has no asymptotes.

$k'(x) = 3 - 3x^2 = 3(1 - x)(1 + x)$, so there is a local minimum at $(-1, -2)$ and a local maximum at $(1, 2)$. $k''(x) = -6x$, so there is a point of inflection at $(0, 0)$.

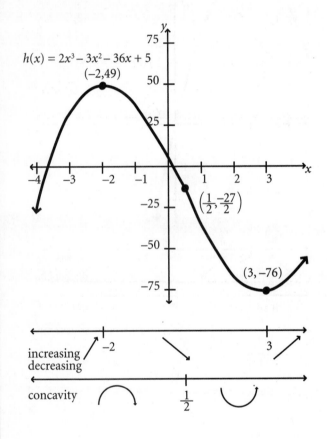

5. $f(x)$ has no asymptotes.
$f'(x) = 4x^3 - 24x^2 = 4x^2(x - 6)$, so there is a local minimum at $(6, -427)$.
$f''(x) = 12x^2 - 48x = 12x(x - 4)$, so there are points of inflection at $(0, -5)$ and at $(4, -251)$.

6. $g(x)$ has a vertical asymptote at $x = -2$ and a horizontal asymptote at $y = 1$. The first derivative is $g'(x) = \dfrac{2}{(x + 2)^2}$, and the second is $g''(x) = \dfrac{-4}{(x + 2)^3}$.

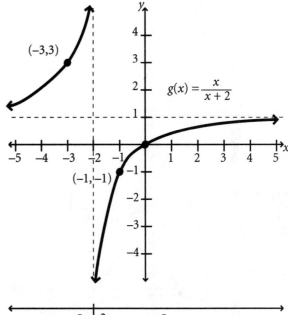

7. $h(x) = \dfrac{1}{(x-3)(x+3)}$ has vertical asymptotes at $x = -3$ and $x = 3$, and a horizontal asymptote at $y = 0$. Because $h'(x) = \dfrac{-2x}{(x^2-9)^2} = \dfrac{-2x}{(x-3)^2(x+3)^2}$, there is a local maximum at $\left(0, -\dfrac{1}{9}\right)$. The second derivative is $h''(x) = \dfrac{6x^2+18}{(x^2-9)^3} = \dfrac{6x^2+18}{(x-3)^3(x+3)^3}$.

8. $k(x) = \dfrac{x}{(x-1)(x+1)}$ has vertical asymptotes at $x = 1$ and $x = -1$, and a horizontal asymptote at $y = 0$.
The first derivative is $k'(x) = -\dfrac{x^2+1}{(x^2-1)^2} = -\dfrac{x^2+1}{(x-1)^2(x+1)^2}$, and the second derivative is $k''(x) = \dfrac{2x(x^2+3)}{(x-1)^3(x+1)^3}$. There is a point of inflection at $(0,0)$.

9. $j(x)$ has a vertical asymptote at $x = 0$ but no horizontal asymptote. Because $j'(x) = \dfrac{x^2 - 1}{x^2} = \dfrac{(x - 1)(x + 1)}{x^2}$, there is a local maximum at $(-1, -2)$ and a local minimum at $(1, 2)$. The second derivative is $j''(x) = \dfrac{2}{x^3}$.

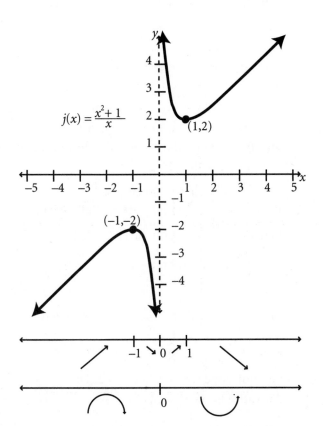

10. $f(x)$ has a horizontal asymptote at $y = 0$ but no vertical asymptotes. Because $f'(x) = \dfrac{1 - x^2}{(x^2 + 1)^2} = \dfrac{(1 - x)(1 + x)}{(x^2 + 1)^2}$, there is a local minimum at $\left(-1, -\dfrac{1}{2}\right)$ and a local maximum at $\left(1, \dfrac{1}{2}\right)$. Because

$$f''(x) = \frac{2x(x^2 - 3)}{(x^2 + 1)^3}$$

$$= \frac{2x(x - \sqrt{3})(x + \sqrt{3})}{(x^2 + 1)^3},$$ there are points

of inflection at $\left(\sqrt{3}, \dfrac{\sqrt{3}}{4}\right)$, $(0,0)$, and $\left(\sqrt{3}, \dfrac{\sqrt{3}}{4}\right)$.

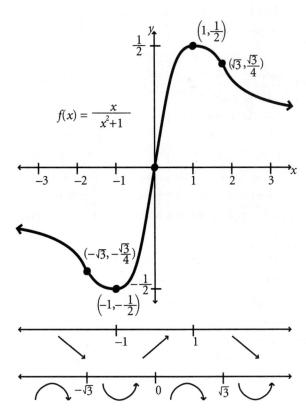

Lesson 15

1. $P'(x) = -\dfrac{1,000}{x^2} + \dfrac{10,000}{x^3} = 0$ when $x = 10$.
Using the first derivative test, $P'(1) = 9,000$ is positive, so the function increases to $x = 10$ and $P'(100) = -\dfrac{1}{10} + \dfrac{1}{100} = -\dfrac{9}{100}$, so the function decreases afterward, thus $x = 10$ maximizes the profit.

2. If x is the number of trees beyond 30 that are planted on the acre, then the number of oranges produced will be:

$O(x) = $ (number of trees) (yield per tree)
$= (30 + x)(500 - 10x) = 15,000 + 200x - 10x^2$.

The derivative $O'(x) = 200 - 20x$ is zero when $x = 10$. Using the second derivative test, $O''(x) = -20$ is negative, so this is maximal. Thus, $x = 10$ more than 30 trees should be planted, for a total of 40 trees per acre.

3. The total sales will be figured as follows:
Sales = (number of copies)(price per copy), so $S(x) = (20 + x)(100 - x) = 2,000 + 80x - x^2$ where x is the number of copies beyond 20. The derivative $S'(x) = 80 - 2x$ is zero when $x = 40$. Because $S''(x) = -2$, this is maximal by the second derivative test. Thus, the artist should make $x = 40$ more than 20 paintings, for a total of 60 paintings in order to maximize sales.

4. After x days, there will be $200 + 5x$ pounds of watermelon, which will valued at $90 - x$ cents per pound. Thus, the price after x days will be $P(x) = (200 + 5x)(90 - x) = 18,000 + 250x - 5x^2$ cents. The derivative is $P(x) = 250 - 10x$, which is zero when $x = 25$. Because $P(x)$ is clearly positive when x is less than 25 and negative afterward, this is maximal by the first derivative test. Thus, the watermelons will fetch the highest price in 25 days.

5. The area is Area $= xy$ and the total fencing is $4y + 2x = 400$. Thus, $x = 200 - 2y$, so the area function can be written as follows:
$A(y) = xy = (200 - 26) \cdot y = 200y - 2y^2$. The derivative $A'(y) = 200 - 4y$ is zero when $y = 50$. Because the second derivative is $A''(y) = -4$, this is an absolute maximum. Thus, the optimal dimensions for the pen are $y = 50$ feet and $x = 200 - 2y = 200 - 2(50) = 100$ feet.

6. Here, Area $= xy$ and the total fencing is $5y + x = 150$. Because $x = 150 - 5y$, the area function can be written as follows:
$A(y) = (150 - 5y)y = 150y - 5y^2$. The derivative $A'(y) = 150 - 10y$ is zero when $y = 15$. Because $A''(y) = -10$, this is an absolute maximum. Thus, the optimal dimensions are $y = 15$ feet and therefore $x = 150 - 5(15) = 75$ feet.

7. Because Volume $= \pi r^2 h = 16\pi$, it follows that $h = \dfrac{16}{r^2}$. Thus, the surface area function is

$$A(r) = 2\pi r^2 + 2\pi r\left(\dfrac{16}{r^2}\right) = 2\pi r^2 + \dfrac{32\pi}{r}$$

The derivative $V'(x) = 150 - \dfrac{3}{2}x^2$ is zero when $4\pi r = \dfrac{32\pi}{r^2}$, so $r^3 = 8$. Thus, the only point of slope zero is when $r = 2$. The second derivative is $A''(r) = 4\pi + \dfrac{64\pi}{r^3}$, which is positive when $r = 2$. Thus, by the second derivative test, $r = 2$ is the absolute minimum. Thus, a radius of $r = 2$ inches and a height of $h = \dfrac{16}{r^2} = 4$ inches will minimize the surface area.

8. Because the box has a square bottom, let x be both its length and its width, and let y be the height. Thus, the volume is Volume $= x^2 y$ and the surface area is Area $= x^2 + 4xy + x^2$ (the top, the four sides, and the bottom). And because Area $= 2x^2 + 4xy = 600$, the height

$$y = \frac{600 - 2x^2}{4x} = \frac{150}{x} - \frac{x}{2}.$$ Thus, the volume is

$$V(x) = x^2\left(\frac{150}{x} - \frac{x}{2}\right) = 150x - \frac{1}{2}x^3.$$

The derivative $V'(x) = 150 - \frac{3}{2}x^2$ is zero when $x^2 = 100$. Negative lengths are impossible, therefore this is zero only when $x = 10$. By the second derivative test, $V''(x) = -3x$ is negative when $x = 10$, so this is a maximum. The corresponding height is $y = \frac{150}{10} - \frac{10}{2} = 10$ feet, so the largest box is a cube with all sides of length 10 feet.

9. Because the box has a square bottom, let x be both the length and width, and y be the height. The area of each side is thus xy, so the cost to build four of them at ten cents a square foot is $0.10(4xy) = 0.4xy$ dollars. The area of the top is $x \cdot x = x^2$, so it will cost x^2 dollars to build. Similarly, it will cost $7x^2$ dollars to build the base. The total cost of the box is therefore Cost $= 0.4xy + 8x^2$. Because the volume is $x^2 y = 40{,}000$, we know $y = \frac{40{,}000}{x^2}$. Thus, the cost function can be written:

$$C(x) = 0.4x\left(\frac{40{,}000}{x^2}\right) + 8x^2$$

$$= \frac{16{,}000}{x} + 8x^2.$$

The derivative:

$$C'(x) = -\left(\frac{16{,}000}{x^2}\right) + 16x$$

is zero only when $x^3 = 1{,}000$ or $x = 10$. By the second derivative test,

$$C''(x) = \left(\frac{32{,}000}{x^3}\right) + 16$$ is positive when

$x = 10$, so this is the absolute minimum. The cheapest box will be built when $x = 10$ feet and $y = 400$ feet.

10. Inside the margins, the area is:

$$\text{Area} = \left(x - 2\left(\frac{3}{2}\right)\right)(y - 2)$$

$$= (x - 3)(y - 2) = xy - 2x - 3y + 6.$$

The total area of the page is $xy = 96$, so $y = \dfrac{96}{x}$. Therefore, the area is

$$A(x) = x\left(\frac{96}{x}\right) - 2x - 3\left(\frac{96}{x}\right) + 6$$

$$= 102 - 2x - \frac{288}{x}.$$

The derivative $A'(x) = -2 + \dfrac{288}{x^2}$ is zero when $x^2 = 144$, thus when $x = 12$ ignore negative lengths. The second derivative $A''(x) = -\dfrac{576}{x^3}$ is negative when $x = 12$, so this is the absolute maximum. Thus, the dimensions that maximize the printed area are $x = 12$ inches and $y = 8$ inches.

12. $-\dfrac{1}{2}$

13. 16

14. 6

15. 0

16. $\dfrac{21}{2}$

17. 35

18. 48

19. 5

20. -3

21. 7

22. 7

23. -11

24. -1

25. 8

26. -17

27. 25

28. 11

29. -1

30. 1

Lesson 16

1. 2

2. $\dfrac{3}{2}$

3. $\dfrac{7}{2}$

4. 2π

5. 3

6. $3 + 2\pi$

7. $-\dfrac{3}{2}$

8. 0

9. $-\dfrac{3}{2}$

10. 4

11. 0

Lesson 17

1. 0 (Note that $\displaystyle\int_a^a g(x)dx = 0$, for any constant a.)

2. $\dfrac{5}{2}$

3. 8

4. $\dfrac{33}{2}$

5. 28

6. $\dfrac{85}{2}$

7. 0

8. 7

9. 14

10. 21

11. 28

12. 35

13. 10

14. -4

15. 36

16. 20

17. $\dfrac{2}{3}$

18. $\dfrac{16}{3}$

19. $\dfrac{38}{3}$

20. $\dfrac{2,000}{3}$

21. $\dfrac{\sqrt{2}}{2}$

22. $\dfrac{1}{2}$

23. $-\dfrac{\sqrt{2}}{2}$

24. 1

Lesson 18

1. $\dfrac{1}{5}x^5 + c$

2. $\dfrac{1}{13}x^{13} + c$

3. $\dfrac{1}{7}u^7 + c$

4. $\left[\dfrac{1}{3}x^3\right]_0^6 = 72$

5. 0

6. $\dfrac{1}{-2}t^{-2} + c = -\dfrac{1}{2t^2} + c$

7. $\left[-t^{-1}\right]_{-4}^{-1} = \left[-\dfrac{1}{t}\right]_{-4}^{-1} = \dfrac{3}{4}$

8. $\dfrac{3}{8}x^{\frac{8}{3}} + c$

9. $\dfrac{4}{5}x^{\frac{5}{4}} + c$

10. $\dfrac{3}{4}u^{\frac{4}{3}} + c$

11. $5x + c$

12. 40

13. $\dfrac{9}{5}x^5 + c$

14. $\dfrac{8}{3}u^3 + c$

15. $\dfrac{3}{4}x^{\frac{4}{3}} - \dfrac{2}{3}x^{\frac{3}{2}} + c$

16. $2x^3 - 5x^2 + 5x + c$

17. 48

18. 11

19. $\dfrac{1}{4}t^{12} + 3t^3 + \dfrac{1}{2}t^2 + c$

20. -6

21. $-\dfrac{1}{2}t^{-4} + \dfrac{4}{3}t^{-3} - \dfrac{1}{2}t^2 + c$

22. 58

23. 208

24. $\dfrac{9}{13}x^{\frac{13}{3}} - \dfrac{56}{11}x^{\frac{11}{7}} + c$

25. $\dfrac{1}{3}x^3 - 5\sin(x) + c$

26. $3e^x + \dfrac{1}{2}x^4 + c$

27. $\left[2\ln|u|\right]_{e}^{e^{2}} = 2\ln\left|e^{2}\right| - 2\ln|e|$

$\qquad\qquad = 4\ln(e) - 2\ln(e) = 4 - 2 = 2$

28. $\frac{1}{2}\theta^{2} - 2\cos(\theta) + c$

29. $-5\cos(x) + 2e^{x} + c$

30. $\left(\frac{1}{2} + e^{1}\right) - (0 + e^{0}) = e - \frac{1}{2}$

31. $\left[4e^{x}\right]_{\ln(2)}^{\ln(3)} = 4e^{\ln(3)} - 4e^{\ln(2)} = 4(3) - 4(2) = 4$

32. $\left[8\sin(x)\right]_{\frac{\pi}{6}}^{\frac{5\pi}{6}} = 0$

Lesson 19

1. $\frac{1}{40}(x^{5} + 1)^{8} + c$, by substituting $u = x^{5} + 1$

2. $\frac{1}{44}(4x + 3)^{11} + c$, by substituting $u = 4x + 3$

3. $\frac{1}{15}$, by substituting $u = x^{3} + 1$

4. $\frac{1}{4}x^{4} - \frac{9}{2}x^{2} + 4x + c$

5. $\frac{3}{8}(x^{2} - 1)^{\frac{4}{3}} + c$, by substituting $u = x^{2} - 1$

6. $\frac{1}{3}(2x + 1)^{\frac{3}{2}} + c$, by substituting $u = 2x + 1$

7. $\frac{5}{6}$, by substituting $u = 1 - x$

8. $\ln\left|3x^{3} - 5x\right|$, by substituting $u = 3x^{3} - 5x$

9. $\frac{1}{2}\sin(x^{4}) + c$, by substituting $u = x^{4}$

10. $\sqrt{3x^{4} - 2x + 1} + c$, by substituting $u = 3x^{4} - 2x + 1$

11. $\frac{1}{4}(4x^{2} + 5x - 1)^{4} + c$, by substituting $u = 4x^{2} + 5x - 1$

12. $-\dfrac{1}{16(4x^{2} + 5)^{2}} + c$, by substituting $u = 4x^{2} + 5$

13. $\frac{1}{2}\ln|4x + 10| + c$, by substituting $u = 4x + 10$

14. Using $u = \sin(x)$, the solution is

$\frac{1}{2}\sin^{2}(x) + c$. Using $u = \cos(x)$, the solution is $-\frac{1}{2}\cos^{2}(x) + c$. Because

$\sin^{2}(x) + \cos^{2}(x) = 1$, these solutions will be the same if the second $+c$ is $\frac{1}{2}$ greater than the first one.

15. $\frac{1}{3}\sin^{3}(x) + c$, by substituting $u = \sin(x)$

16. $\frac{1}{4}\sin(4x) + c$, by substituting $u = 4x$

17. $4\sin(x) + c$

18. $-\frac{1}{7}\cos(7x - 2) + c$, by substituting $u = 7x - 2$

19. $\cos(e^{x}) + c$, by substituting $u = e^{x}$ (the one inside $\sin(e^{x})$)

20. $\frac{1}{4}(\ln(x))^{4} + c$, by substituting $u = \ln(x)$

21. $\ln|\ln(x)| + c$, by substituting $u = \ln(x)$

22. $2e^{\sqrt{x}} + c$, by substituting $u = \sqrt{x}$ (the one occurring in the term $e^{\sqrt{x}}$)

23. $-\ln|\cos(x)| + c$, by substituting $u = \cos(x)$

24. $\frac{1}{2}\ln(1 + e^{2x}) + c$, by substituting $u = 1 + e^{2x}$

Lesson 20

1. $\frac{1}{6}x^6\ln(x) - \frac{1}{36}x^6 + c$, done by parts with $u = \ln(x)$

2. $-x\cos(x) + \sin(x) + c$, by parts with $u = x$

3. $-\frac{1}{2}\cos(x^2) + c$, by the substitution $u = x^2$

4. $(x + 3)\sin(x) + \cos(x) + c$, by parts with $u = x + 3$

5. $\frac{1}{2}(\ln(x))^2 + c$, by substituting $u = \ln(x)$

6. $-x^2\cos(x) + 2x\sin(x) + 2\cos(x) + c$, using parts twice, first with $u = x^2$, and then with $u = 2x$

7. $\frac{1}{3}x^3 - \cos(x) + c$, by basic integration

8. $\frac{1}{3}e^{x^3+1} + c$, by substituting $u = x^3 + 1$

9. $\frac{1}{3}\ln(e^{3x} + 9) + c$, by substituting $u = e^{3x} + 9$

10. $\left(\frac{1}{4}x^4 + \frac{3}{2}x^2 - x\right)\ln(x) - \frac{1}{16}x^4 - \frac{3}{4}x^2 + x + c$, by parts with $u = \ln(x)$

11. $\ln|x| + x\ln(x) - x + c$, evaluating $\int \ln(x)\,dx$ by parts with $u = \ln(x)$

12. $\frac{2}{3}(x - 1)^{\frac{3}{2}} + c$, substituting $u = x - 1$

13. $\frac{2}{3}x(x - 1)^{\frac{3}{2}} - \frac{4}{15}(x - 1)^{\frac{5}{2}} + c$, by parts with $u = x$

14. $-xe^{-x} - e^{-x} + c$, by parts with $u = x$

15. $-\frac{1}{4}\cos^4(x) + c$, by substituting $u = \cos(x)$

16. $-\frac{1}{4(\ln(x))^4} + c$, by substituting $u = \ln(x)$

17. $\frac{1}{3}\ln|x| + c$, by basic integration

18. $e^{-\cos(x)} + c$, by substituting $u = -\cos(x)$

19. $-e^{\frac{1}{x}} + c$, by substituting $u = \frac{1}{x}$

20. $-\frac{2}{3}(\cos(x))^{\frac{3}{2}} + c$, by substituting $u = \cos(x)$

21. First, let $u = \text{six}(x)$.

Then, $\int \cos(x)\ln(\sin(x))\,dx = \int \ln(u)\,du$.

From an earlier example,

$\int \ln(u)\,du = u\ln(u) - u + c$. So, we get

$\sin(x) \cdot \ln(\sin(x)) - \sin(x) + c$.

22. $\frac{1}{2}(e^x\sin(x) + e^x\cos(x)) + c$, by parts twice, plus the trick from the last example in the lesson.

GLOSSARY

acceleration the rate at which the velocity of a moving object is increasing or decreasing

additive rule parts of a function added together can be differentiated separately:
$$\frac{d}{dx}(f(x) + g(x)) = f'(x) + g'(x)$$

antiderivative given a function $f(x)$, the anti-derivative is a function $g(x)$ such that $g'(x) = f(x)$.

asymptote a line that a graph flattens out toward. It occurs as either an infinite limit or a limit as $x \to \pm\infty$.

Chain Rule $\frac{d}{dx}(f(g(x))) = f'(g(x)) \cdot g'(x)$

closed interval the set of all the real numbers between and including two endpoints, like all x such that $a \le x \le b$

composition the process of plugging one function into another. The composition of functions f and g is $(f \circ g)(x) = f(g(x))$.

concave down when the graph of a function bows downward, like a frown

concave up when the graph of a function bows upward, like a smile

concavity the way a graph curves either upward or downward

Constant Coefficient Rule a constant c multiplied in front of a function is unaffected by differentiation: $\frac{d}{dx}(c \cdot f(x)) = c \cdot f'(x)$

Constant Rule the derivative of a constant is zero.

continuous function one whose graph can be drawn without picking up the pencil

cosecant abbreviated csc; see *trigonometry*

cosine abbreviated cos; see *trigonometry*

cotangent abbreviated cot; see *trigonometry*

critical points points of slope zero, points where the derivative is undefined, and endpoints of the domain

decreasing when the graph of function goes down from left to right

definite integral the area between a graph $y = f(x)$ and the x-axis from $x = a$ to $x = b$ where area below the x-axis counts as negative, written $\int_b^a f(x)dx$

degrees measure the size of angles in such a way that a complete circle is 360°

derivative the derivative of function $y = f(x)$ is
$$\frac{dy}{dx} = f'(x) = \lim_{a \to 0} \frac{f(x + a) - f(x)}{a},$$
which is the slope of the tangent line at point $(x, f(x))$.

discontinuity a break in a graph

domain the set of all the real numbers at which a function can be evaluated

e a transcendental number approximately equal to $2.71828\ldots$

explicit a function is explicit if its formula is known exactly.

exponent an exponent says how many times a factor is multiplied by itself. In the case of roots, the exponent is a fraction.

First Derivative Test if a function increases to a point and then decreases afterward, then that point is a local maximum. If the function decreases to a point and then increases afterward, then the point is a local minimum.

function a mathematical object that assigns one number in its range to every number in its domain

Fundamental Theorem of Calculus if
$$g(x) = \int_0^x f(t)dt, \text{ then } g'(x) = f(x). \text{ Thus,}$$
$$\int_a^b f(x)dx = g(b) - g(a) \text{ where } g'(x) = f(x).$$

graph a visual depiction of a function where the height of each point is the value assigned to the number on the horizontal axis

horizontal asymptote a horizontal line toward which the graph flattens out as $x \to \infty$ or $x \to -\infty$

implicit a function is implicit if it was defined in an indirect manner so that its exact formula is unknown.

implicit differentiation the process of taking a derivative of both sides of an equation and using the Chain Rule with $\frac{d}{dx}(x) = 1$, $\frac{d}{dx}(y) = \frac{dy}{dx}$, $\frac{d}{dt}(x) = \frac{dx}{dt}$, and so on

increasing when the graph of a function goes up from left to right

indefinite integral represents the antiderivative:
$$\int f(x)dx = g(x) + c \text{ if and only if } g'(x) = f(x)$$

integral see either *definite integral* or *indefinite integral*

L'Hôpital's Rule If $\lim_{x \to \infty} f(x) = \pm\infty$ and $\lim_{x \to \infty} g(x) = \pm\infty$, then $\lim_{x \to \infty} \frac{f(x)}{g(x)} = \lim_{x \to \infty} \frac{f'(x)}{g'(x)}$. The same is true when $\lim_{x \to -\infty}$.

limit the limit $\lim_{x \to a} f(x) = L$ means that the values of $f(x)$ get very close to L as x gets close to a.

limit from the left $\lim_{x \to a^-} f(x) = L$ means that the values of $f(x)$ are close to L when x is close to, and less than, a.

limit from the right $\lim_{x \to a^+} f(x) = L$ means that the values of $f(x)$ are close to L when x is close to, and greater than, a.

limits at infinity $\lim_{x \to \infty} f(x) = L$ means that the values of $y = f(x)$ get close to $y = L$ as x gets really big. If large negative values of x are used and $y = f(x)$ gets close to $y = L$, then $\lim_{x \to -\infty} f(x) = L$.

limits of integration the limits of the integral $\int_a^b f(x)dx$ are a and b.

local maximum the highest point on a graph in that immediate area, like a hilltop

local minimum   the lowest point on a graph in that immediate area, like a valley

natural logarithm   the inverse $\ln(x)$ of the exponential function e^x. Thus, $y = \ln(x)$ if and only if $e^y = x$.

oscillate   to repeatedly go back and forth across a range of values

point of inflection   a point on a graph where the concavity changes

point-slope formula   the equation of a straight line through (x_1, y_1) with slope m is
$$y = m(x - x_1) + y_1.$$

polynomial   the sum of powers of a variable, complete with constant coefficients. For example, $x^2 + 3x - 5$ and $10x^7 - 12x^5 + 4x^2 - x$ are both polynomials.

position function   gives the mark on a line where a moving object is at a given time

Power Rule $\dfrac{d}{dx}(x^n) = n \cdot x^{n-1}$

Product Rule
$$\frac{d}{dx}(f(x) \cdot g(x)) = f'(x) \cdot g(x) + g'(x) \cdot f(x)$$

Pythagorean theorem   the squares of the legs of a right triangle add up to the square of the hypotenuse.

Quotient Rule
$$\frac{d}{dx}\left(\frac{f(x)}{g(x)}\right) = \frac{f'(x) \cdot g(x) - g'(x) \cdot f(x)}{(g(x))^2}$$

radians   measure the size of angles in such a way that a complete circle is 2π radians

range   the set of all the numbers that can be the value of a function

rate of change   how fast a quantity is increasing or decreasing

secant   abbreviated sec; see *trigonometry*

rational function   a rational function is the quotient of two polynomials. For example, $\dfrac{8x^3 - 10x + 4}{5x - 2}$ is a rational function.

second derivative   the derivative of the first derivative

Second Derivative Test   a point of slope zero is the maximum if the second derivative is negative at the point, and a minimum if the second derivative is positive at the point.

sign diagram   tells where an expression is positive and negative

sine   abbreviated sin; see *trigonometry*

slope   a measure of steepness of a straight line. It is the amount the y-value goes up or down with each step to the right.

slope-intercept formula   the equation of a straight line with slope m that crosses the y-axis at $y = b$ is
$$y = mx + b.$$

Squeeze Theorem   if $f(x) \leq g(x) \leq h(x)$ and $\lim\limits_{x \to a} f(x) = L = \lim\limits_{x \to a} h(x)$, then $\lim\limits_{x \to a} g(x) = L$.

substitution   an integration technique used to reverse the Chain Rule

tangent   abbreviated tan; see *trigonometry*

tangent line   a straight line that indicates the direction of a curve at a given point

third derivative   the derivative of the second derivative

trigonometric identities $\tan(x) = \dfrac{\sin(x)}{\cos(x)}$,

$\sec(x) = \dfrac{1}{\cos(x)}$, $\csc(x) = \dfrac{1}{\sin(x)}$,

$\cot(x) = \dfrac{\cos(x)}{\sin(x)}$, and $\sin^2(x) + \cos^2(x) = 1$

trigonometry the study of functions formed by dividing one side of a right triangle by another. When the right triangle has angle x, the hypotenuse has length H, the side adjacent to x has length A, and the side opposite x has length O, then

$$\sin(x) = \frac{O}{H}$$

$$\cos(x) = \frac{A}{H}$$

$$\sec(x) = \frac{H}{A}$$

$$\csc(x) = \frac{H}{O}$$

$$\tan(x) = \frac{O}{A}$$

$$\cot(x) = \frac{A}{O}$$

The derivatives are:

$$\frac{d}{dx}(\sin(x)) = \cos(x)$$

$$\frac{d}{dx}(\cos(x)) = -\sin(x)$$

$$\frac{d}{dx}(\sec(x)) = \sec(x)\tan(x)$$

$$\frac{d}{dx}(\csc(x)) = -\csc(x)\cot(x)$$

$$\frac{d}{dx}(\tan(x)) = \sec^2(x)$$

$$\frac{d}{dx}(\cot(x)) = -\csc^2(x)$$

unit circle the circle of radius 1 centered at the origin

velocity the rate of change of a moving object at a particular time

vertical asymptote a vertical line $x = a$ that a graph more closely resembles as the inputs x approach a

ADDITIONAL ONLINE PRACTICE

Whether you need help building basic skills or preparing for an exam, visit LearningExpress Practice Center! On this site, you can access additional practice materials. Using the code below, you'll be able to log in and access a practice Calculus Success in 20 Minutes a Day exam. This online practice will also provide you with:

Immediate scoring
Detailed answer explanations
A customized diagnostic report that will assess your skills and focus your study

Log in to the LearningExpress Practice Center by using the URL: www.learnatest.com/practice

This is your Access Code: **8899**

Follow the steps online to redeem your access code. After you've used your access code to register with the site, you will be prompted to create a username and password. For easy reference, record them here:

Username: _____   Password: _____

With your username and password, you can log in and access your additional practice exams. If you have any questions or problems, please contact LearningExpress customer service at 1-800-295-9556 ext. 2, or e-mail us at **customerservice@learningexpressllc.com**